X-ray and Neutron Diffraction

X-ray and Neutron Diffraction

by

G. E. BACON, M.A., Sc.D.

Professor of Physics, University of Sheffield

PERGAMON PRESS

OXFORD · LONDON · EDINBURGH · NEW YORK
TORONTO · SYDNEY · PARIS · BRAUNSCHWEIG

Pergamon Press Ltd., Headington Hill Hall, Oxford
4 & 5 Fitzroy Square, London W.1

Pergamon Press (Scotland) Ltd., 2 & 3 Teviot Place, Edinburgh 1

Pergamon Press Inc., 44–01 21st Street, Long Island City, New York 11101

Pergamon of Canada, Ltd., 6 Adelaide Street East, Toronto, Ontario

Pergamon Press (Aust.) Pty. Ltd., 20–22 Margaret Street, Sydney, New South Wales

Pergamon Press S.A.R.L., 24 rue des Écoles, Paris 5ᵉ

Vieweg & Sohn GmbH, Burgplatz 1, Braunschweig

Copyright © 1966 Pergamon Press Ltd.
First edition 1966
Library of Congress Catalog Card No. 66–25305

Printed in Great Britain by Bell and Bain, Ltd., Glasgow

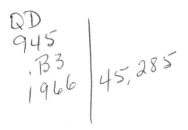
(2957/66)

Contents

Preface

THE study of solids by diffraction techniques may be regarded as the twentieth-century sequel to classical crystallography and the culmination of the work and ideas of crystallographers during the previous 200 years. So far as X-rays are concerned the starting date is 1912 and the technique is just over fifty years old. Some of the first workers in the subject are still actively engaged in it today, but, as present-day techniques go, it would not be considered young.

Over the fifty years it is possible to see a clear pattern of development and it is the aim of the present book to trace the growth of the subject, illustrating this with a selection of original papers. In attempting this task the writer immediately disclaims two things. He seeks to establish no historical, or national, order of precedence for any of the work and ideas discussed, nor does he suggest that the papers chosen are the most significant or most important ones. They have been chosen simply to illustrate conveniently the writer's view of how diffraction analysis has developed, a story told by someone whose introduction to this field of scientific research came exactly half-way through its present span of fifty-four years.

During the latter twenty years X-ray diffraction has been complemented and extended by the use of neutron diffraction, a technique with immense scope but suffering, from the practical point of view, the great limitation of requiring a nuclear reactor as the source of neutrons. The reprinted papers are divided roughly equally between X-rays and neutrons. The nine chapters which comprise the first section of the book discuss the principles and uses of X-rays and neutrons as they have developed from the foundation provided by classical crystallography. An outline

A•

is given of the formal classification of extended arrangements into Space Groups as an essential first step in using diffraction methods to discover unknown structures. If the reader has any feeling that either this aspect of the subject or the discussion of crystal form is overweighted, then the writer will claim that their value is usually underestimated. The reprint section includes twenty papers, mainly in full, to represent the development of the techniques and they are distributed fairly evenly in time among the intervening years. One further short paper is incorporated in the text of Chapter VII. The writer is grateful to the authors and the publishers of the journals in which these articles appeared for permission to reprint them. In particular thanks are due to the Editor of *Physical Review* for the papers numbered 4, 10, 12, 13, 14, 17, 18, 19 and 20, to the Royal Society for 5 and 15, to the Institute of Physics and the Physical Society for 7 and 8, to Messrs. Gauthier-Villars for 11 and 21, to the American Institute of Physics for 16, to the Cambridge Philosophical Society for 2, to the Chemical Society for 9 and to the publishers of the *Proceedings of the Bavarian Acadamy of Sciences*, the *Philosophical Magazine* and the *Zeitschrift für Kristallographie* for 1, 3 and 6, respectively.

Although the chapters in Part I are in a sense a commentary on the papers which follow, it will be found that they are very largely self-contained and can be read sensibly alone. It is suggested that the reader, who has the time, will get most benefit from the book by a first reading from beginning to end, followed by a return to the introduction and a more detailed study of individual papers.

PART 1

Commentary

CHAPTER I

Crystal Form and Crystal Symmetry

As a result of the experiments of physicists and chemists it is now an accepted fact that matter of all kinds, whether it be in the solid, liquid or gaseous form, is built up out of atoms. For over fifty years it has been known that in the most highly developed form of matter, namely in solids, the arrangement of the atoms is extremely regular and forms a continuous pattern in three dimensions in space. The purpose of this book is to describe how this highly regular arrangement can be determined in detail by diffraction methods. Various radiations can be used for this purpose, in particular X-rays, neutrons and electrons, and one of our purposes will be to see the limitations and advantages of the individual techniques. We shall find that solids are built up of regular building blocks or "unit cells" as we shall call them. Each unit cell may consist of only a few atoms but by continued stacking of the cells we arrive at a sizeable, visible piece of solid matter. Our first aim will be to determine the size and shape of these unit cells and then to find out the distribution of the atoms within them. This will give us a preliminary picture of the solid but it will only be a static picture, that is a picture in which the atoms are stationary, and this we shall have to develop further by adding details to describe the way in which the atoms move about because of the thermal energy which they possess. For solids the motion is restricted to oscillation about a mean equilibrium position and we shall find that this movement is of two kinds. First there is random motion of individual atoms, and secondly there is connected motion between linked groups of atoms, such as rotation and vibration of whole molecules or atomic groups. A further detail, which we shall have to add to our picture in certain cases, is information which will enable us to describe the magnetic properties of the solid. Atoms of certain materials are magnetic in the sense that they behave as though they carry minute magnetic-compass needles. In particular we shall show how to find the magnitude of the magnetic moments of these minute compass needles and also to work out the directions in which they point, or, as we usually describe it, their orientation. At the outset our pictures of static atoms in crystals will be related to the ideal case, where the three-dimensional order is perfect and there are no mistakes. Later on we shall see that most crystals are not entirely perfect but have defects in them. As a more extreme example of this, where departures from perfection become increasingly great, we change from the case of a solid to a liquid.

In general our plan will be to follow up the phases of development of the subject as they have occurred historically. First of all we shall look at the morphology of crystals, that is the arrangement of crystal faces and the details of crystal geometry. We shall see that crystals have anisotropic properties, which means that some of their physical properties differ in different directions, and this is a direct outcome of the anisotropy of their atomic structure. We shall then follow the development of methods of deducing the arrangement of atoms by using X-rays, employing the technique of X-ray diffraction, which at least in favourable cases leads us directly to the details of the three-dimensional structure. Finally, we shall see how this picture has been developed further by using neutrons as the bombarding radiation rather than X-rays, and it is this use of the technique of neutron diffraction which will reveal to us the "magnetic architecture" of those solids which contain magnetic atoms.

In a study of solids it is natural to begin with the best developed form of a solid, namely the crystal. All solids are crystalline to a very large degree, and although most solids may appear very imperfect on a macroscopic scale, and quite different from the large crystals which we can observe in the mineralogical collections of museums, nevertheless they are almost invariably built up of crystalline regions. In an ordinary piece of metal, for example, these crystalline regions may be too small to be seen with the naked eye, but they can be distinguished in a microscope. Crystals have always fascinated people and scientists have studied them in detail for several centuries. For example, Steno, a Danish geologist, was studying quartz in 1669 and a century later Romé

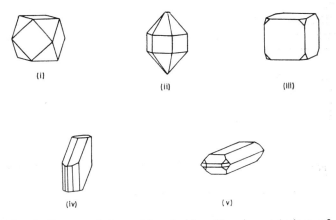

Fig. 1. Some typical crystals, showing (i) cubo-octahedron of fluorspar, (ii) a hexagonal prism of quartz terminated by hexagonal pyramids, (iii) a cube with small secondary octahedral faces, (iv) a crystal of gypsum, and (v) barytes.

de l'Isle and Hauy were making comprehensive measurements of the geometry of a wide variety of crystals. Good specimens of crystals have highly perfect faces, and these faces occur in characteristic angular positions. Usually the faces are very unevenly developed, so that it is difficult at first sight to visualize the underlying symmetry of the crystal form. The actual sizes of the individual faces will depend on chance circumstances which determine the way in which the crystal has grown in different directions, but when idealized the crystals will have regular shapes like those which are shown in Fig. 1. Each of the examples which we see in this figure is a combination of a number of simpler individual forms. For example we may have a combination of an octahedron with a cube, as in (i), (iii), or a pyramid with a vertical prism as in (ii). Or a particular crystal may include several different prisms and pyramids among its faces as in (iv), (v) of Fig. 1. If we measure the angular inclinations of the various faces which occur for a particular species, then we shall find that they occur· at very special angles, and certainly not in random positions. The situation is reminiscent of what we see in a well-planted orchard of fruit trees, where we notice that only at a few special viewpoints do the trees fall into lines. If we look in more

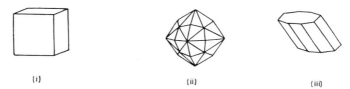

(i)   (ii)   (iii)

FIG. 2. The highly symmetrical forms of (i) the cube and (ii) the hexakisoctahedron in the cubic system are contrasted with the much lower symmetry of a crystal (iii) in the monoclinic system.

detail at a collection of crystal models we shall find that they include representatives of many different types of symmetry. For example, there is the highly developed symmetry of the cubic system as illustrated at (i), (ii) of Fig. 2, and crystals which occur in this class are often metals or simple ionic compounds. A much lower type of symmetry such as (iii) in Fig. 2 which belongs to the monoclinic system, and which we cannot describe in terms of three mutually perpendicular co-ordinate axes, is found very generally among organic materials. The comparison of the cubic and monoclinic systems, as represented by metals and organic molecules respectively, illustrates an important conception. When we are packing together identical spherical atoms or atoms of similar size we are likely to get a highly symmetrical structure. On the other hand, when we are packing together anisotropic shapes, such as the flat molecules which occur rather generally among organic compounds, then the symmetry which results is likely to be very much lower. We may compare the difference in the symmetries of the solids which occur in these two cases with the difference, on a much larger scale, between the regular packing which exists in a pile of ball-bearings and the stacked arrangement which is produced when we try to pack together a collection of cards.

As a result of the extended study of the external forms of crystals, crystallographers developed methods of formal classification, and some acquaintance with these will be necessary if we are to understand the underlying internal structure. We can idealize our crystals by representing their faces by a system of normals drawn perpendicular to the faces as shown in Fig. 3 (i), where the normals a, b, c, d are drawn perpendicular to the faces A, B, C, D respectively. Then, as a further refinement we may represent these normals by points such as a, b ,c, d in Fig. 3 (ii) at which they intersect the surface of a sphere. Our problem is then to study the arrangement of such collections of points since the symmetry of this arrangement represents the symmetry of the crystal. In the case of the example shown in Fig. 3 we see that the crystal symmetry comprises a fourfold tetragonal axis, which is vertical. If we had also included in the diagram points to represent the underneath faces of the crystal, which we omitted for clarity, we should see that there is also a horizontal plane of symmetry. The fourfold axis and the plane of symmetry are examples of "symmetry elements". These elements are such that when they "operate" on the collection of points then the collection returns to its initial arrangement. Thus the operation represented by a fourfold axis is a rotation through 90° and it will be seen that when this operation is carried out by a vertical axis on the set of points which are shown in Fig. 3 (ii) then a set in identical positions is produced. Investigation shows that there can only be a limited number of combinations of symmetry elements grouped together in this way and passing through a point, indicated as O, and it is a purely mathematical problem to work out this total number.

In fact there are found to be thirty-two possible combinations of symmetry elements passing through a point and they are

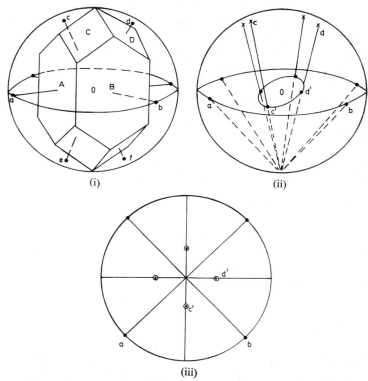

Fig. 3. Representation of the distribution and orientation of crystal faces. In the idealized drawing of a crystal shown in (i) normals are drawn perpendicular to the crystal faces *A*, *B*, *C*, and *D*, to intersect the sphere at *a*, *b*, *c*, *d*, etc. In (ii) these points on the sphere are joined to the south pole of the sphere by lines which intersect the equatorial plane in points such as *c'*, *d'*. These points provide the stereographic representation of the crystal faces, as indicated in (iii) for the crystal initially drawn in (i).

called the "32 point groups". These thirty-two arrangements can be divided into seven symmetry systems, the division being made according to the systems of coordinates to which they can be related. These systems are as follows: (1) the *Cubic* system which has three equal axes at right angles; (2) the *Tetragonal* system, with three axes at right angles, two of them equal and the third one unequal; (3) the *Orthorhombic* system, with three axes at right angles, but all unequal; (4) the *Rhombohedral* system with three equal axes, which are equally inclined but not at right angles; (5) the *Hexagonal* system with three equal co-planar axes inclined at 120° to one another and a fourth unequal axis which is perpendicular to the plane; (6) the *Monoclinic* system with three unequal axes, one of which is perpendicular to the plane of the other two; and (7) the *Triclinic* system with three unequal axes, all inclined at unequal angles which are different from 90°. We can therefore list the parameters which are necessary in order to give a full description of a crystal belonging to each of these systems as follows: for the *cubic* system the length a; for the *tetragonal* system lengths a and c; for the *orthorhombic* system lengths a, b and c; for the *rhombohedral* system length a and the angle α; for the *hexagonal* system length a and length c; for the *monoclinic* system lengths a, b and c and an angle β; for the *triclinic* system lengths a, b and c and three angles α, β and γ.

In order to examine the way in which these thirty-two crystal classes are distributed among the seven systems of crystal symmetry it is convenient to use a method of representing direction which is known as the stereographic projection. In Fig. 3 (ii) we represented each crystal face, or normal direction, by a point such as a, b, c or d on the surface of a sphere. For each of these points, such as c or d, we draw a line to the south pole of the sphere as shown in Fig. 3 (ii) itself. The points, such as c', d' where these lines intersect the equatorial plane of the sphere, are the stereographic representation of the crystal face. Thus Fig. 3 (iii) is the stereogram for the crystal which we first showed in Fig. 3 (i). Points a, b on the circumference of the circle correspond to faces A, B which are vertical in the original crystal. For the crystal faces which are on the underneath of the crystal we shall have to join the ends of their normals, such as e, f in Fig. 3 (i) to

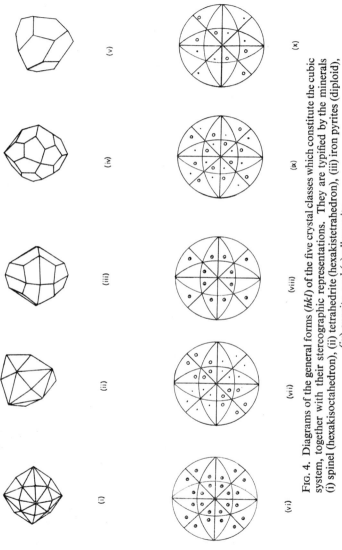

FIG. 4. Diagrams of the general forms (*hkl*) of the five crystal classes which constitute the cubic system, together with their stereographic representations. They are typified by the minerals (i) spinel (hexakisoctahedron), (ii) tetrahedrite (hexakistetrahedron), (iii) iron pyrites (diploid), (iv) cuprite, and (v) ullmannite.

the *north* pole of the sphere. We indicate the existence of these "underneath" normals by the circles which are included in Fig. 3 (iii), overlapping the dots in the stereogram which correspond to the upper faces of the crystal.

Investigation shows that there are five different combinations of symmetry elements which belong to the cubic system. The general form for each of these five classes is shown in Fig. 4 together with the corresponding stereogram. We emphasize that what these stereograms show is for any chosen general direction in space the *other directions* which are identical. We see that in the case of each of the cubic arrangements in Fig. 4 there are four threefold or triad axes. Each arrangement possesses distinctive symmetry elements of its own but they all include four triad axes. Thus the first form which is shown in Fig. 4 is known as the hexakisoctahedron and crystals which belong to this class have four triad axes, three fourfold (tetrad) axes, six twofold (diad) axes, three principal planes of symmetry, six secondary planes and a centre of symmetry. Second is the form characteristic of zinc blende and known as a hexatetrahedron which has four triad axes, three diads, and six planes of symmetry but no centre of symmetry. The third form which we show is characteristic of iron pyrites and has four triad axes, three diads, three planes and again a centre of symmetry.

We can continue in a similar way to subdivide the remaining twenty-seven of the thirty-two crystal classes among the other crystal systems, and to list the primary symmetry elements for each crystal system. We find that the essential elements are as follows: for the cubic system four triad axes, *tetragonal* system one tetrad, *orthorhombic* system three diads at right angles, *rhombohedral* system one triad, *hexagonal* system one hexad, *monoclinic* system one diad, and *triclinic* system no symmetry elements. Figure 5 shows in the form of stereograms the way in which the thirty-two crystal classes have been subdivided into the seven crystal systems on the basis of the symmetry of the external form of the crystals.

We emphasize again that these "crystal classes" are also

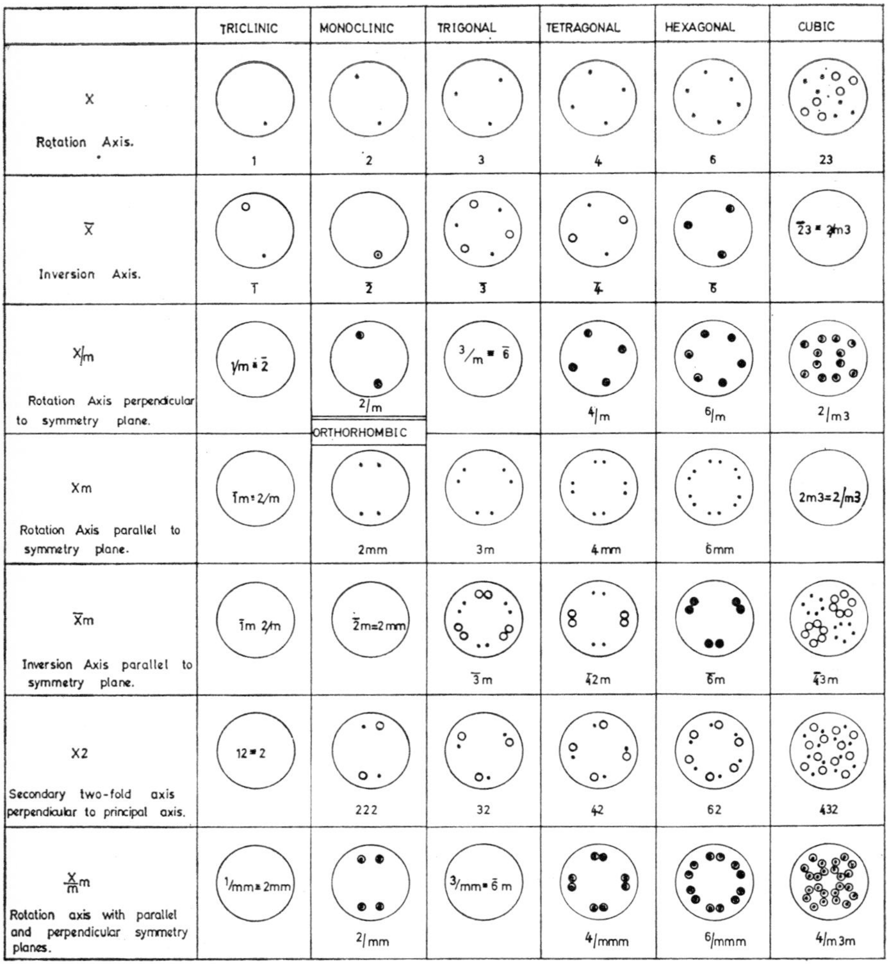

FIG. 5. Stereograms indicating the symmetries of the thirty-two crystal classes, divided according to the crystal systems to which they belong, based on an original diagram by F. C. Phillips.

called "point groups", because they refer to ways of grouping symmetry elements which pass through a point. They describe the grouping of faces on a crystal, or the way in which groups of points derive from a given single point by the operation of symmetry elements. In a first discussion of symmetry elements we think of planes of symmetry, axes of symmetry and a centre of symmetry. Thus a symmetry plane divides an object, or assembly of points, into two sections which are related as object and mirror image. An axis of symmetry, of order n, means that rotation through $360/n$ degrees produces a congruent position. Only values of $n = 2$, 3, 4 and 6 are found and this restriction arises because, as we shall see later, the external form of a crystal reflects its internal structure. This structure is continuous in three dimensions and arises as a result of packing together units of pattern without leaving gaps between them. This can be done with rectangular, trigonal, square and hexagonal prisms but not with any others: thus although hexagons pack together as in a honeycomb it is impossible to pack octagons together without leaving square holes between them. The existence of a centre of symmetry means that every face has a similar face parallel to it as, for example, in an octahedron but not a tetrahedron. Combinations of these elements will produce in fact thirty crystal classes. If, however, we widen the conception of axes to include an inversion axis, which is a combination of rotation and inversion through a centre of symmetry, we then get an additional two possibilities, making thirty-two crystal classes altogether. An "inversion tetrad" axis, for example, implies rotation through an angle of 90°, i.e. 360°/4, followed by inversion through a centre. This in fact gives a new combination, represented by $\bar{4}$ in Fig. 5. The symmetry here is more regular than that shown by a simple diad axis, but less regular than that of a tetrad. In a similar way the symbol $\bar{4}m$ which represents an inversion axis and a symmetry plane parallel to it contributes a new arrangement. On the other hand $\bar{1}$, $\bar{2}$, $\bar{3}$ and $\bar{6}$ contribute nothing new, being equal to a centre, a plane of symmetry, a triad with a centre of symmetry and the combination $3/m$ respectively.

Although we have based our conceptions of symmetry purely on the arrangement of the faces found on crystals, we shall find that similar relationships exist among the physical properties. In general, physical properties of crystals depend on the direction of measurement in the crystal, relative to the crystallographic axes. The symmetry elements of the crystal will determine the equivalence or non-equivalence of different directions in exactly the same way as they relate different crystal faces which are perpendicular to these directions. A full description of the way in which different physical properties vary with direction in the different crystal systems would be much too lengthy to include here and we shall merely take a few examples. Thus, properties such as thermal conductivity require a triaxial ellipsoid for their description, involving nine parameters in the general case. In the tetragonal system the number of parameters has fallen to two and in cubic systems only one parameter remains. Another property which can be described in this way is the velocity of light, in terms of the direction of vibration of its electric field, leading to the discussion of double refraction in crystals and the details of crystal optics. In the extreme case of cubic crystals the transmission of light is isotropic, but cubic crystals are *not* isotropic for all physical properties. Thus the case of elasticity is much more complicated. In general† twenty-one parameters are required to give a full description and even in the cubic system three parameters are needed. This means that the elastic properties of a cubic crystal are not the same in all directions and if we construct surfaces which represent the way in which distortions of various kinds will vary with direction, they will *not* be spheres. We emphasize, however, that although these surfaces are not spheres they will be surfaces which possess the symmetry of the class of the cubic system to which the particular crystal belongs.

---

†Properties such as thermal conductivity and the velocity of light are represented by second order tensors. Elasticity is represented by a fourth-order tensor. Such a tensor can have eighty-one components but many of these may be equal and the elasticity tensor reduces to twenty-one different components in the general case.

They will all possess the minimal cubic symmetry, i.e. four triad axes. Two examples are given for cubic rock-salt in Fig. 6 which shows at (i) the way in which the bending of a plate depends on the direction in which the plate is cut and (ii) the way in which the twist of a cylinder depends on the direction of the cylinder axis. The diagrams indicate planar sections of complicated three-dimensional figures: in each case we would find that the

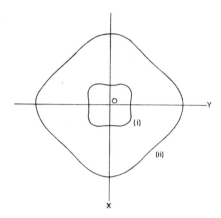

FIG. 6. Planar sections parallel to a cube face of elasticity figures for rock-salt. In (i) the length of any radius vector is proportional to the bending modulus of a plate cut parallel to that direction: in (ii) the radius vector is proportional to the rigidity modulus for the twisting of a cylinder having its axis parallel to that direction.

section of this figure by a plane perpendicular to a triad axis was a circle, so that the equivalence of directions at 120° intervals is certainly preserved.

Let us look at one or two examples of some more unusual physical properties which are found only to occur in certain crystal classes. For example there is a phenomenon of "optical rotation" or "optical activity". It is possible for this to occur, but it is not necessarily present, in classes of symmetry which possess no planes of symmetry. That is, it can be found in the rows X, $\bar{X}$, and $X2$ in Fig. 5. Quartz, which is in class 32, shows

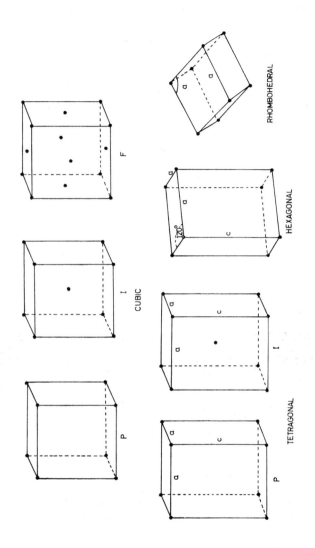

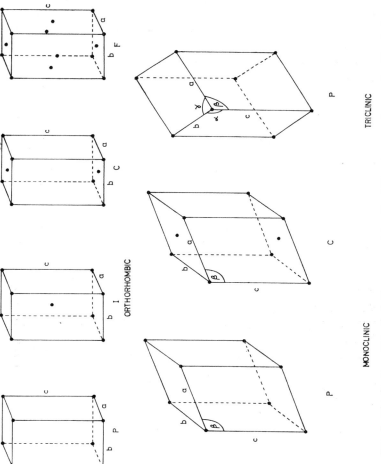

FIG. 7. The fourteen Bravais lattices, comprising the different ways of arranging identical points in space so as to give continuous three-dimensional distributions. The lattices for the monoclinic and triclinic system are viewed in a direction which is perpendicular to the conventional one, in order to indicate more clearly the departures from orthogonality.

optical activity. On the other hand piezo and pyro electricity can be shown by crystals in classes which do not have a centre of symmetry. Thus tourmaline in class $3m$ is a pyro-electric crystal. An example of a piezo-electric crystal is quartz, which, as we have already seen, is in class 32, a class of symmetry which possesses neither a plane nor centre of symmetry.

So far all our thoughts about symmetry have been related to symmetries of isolated groups of points, i.e. to assemblies about a single point in space. Thus, for the crystal forms and stereograms, which we discussed in Fig. 5, for example, all the symmetry planes and axes pass through an individual point. In considering the structure and symmetry of solids we shall find that we have to extend these conceptions in some kind of way which will permit description of continuous distributions or structures in three dimensions. In order to widen our ideas in this way we must first of all find out how many ways there are of assembling an infinite array of points, in such a manner that the environment of each one is the same as that of all the others. There are, perhaps rather surprisingly, only fourteen ways in which this can be done. These fourteen different arrangements are called Bravais lattices. They are shown in Fig. 7. In order to develop a complete system of arrangements in three dimensions, we then have to associate each of our point groups, in turn, with the Bravais lattices which have the appropriate symmetry. Not every combination will be found to produce a unique arrangement but there will obviously be many more combinations than 32. However, before proceeding further in this way let us list the details of the Bravais lattices themselves. In the cubic system there are three: a simple one denoted by P, a body-centred one I and a face-centred one F. In the tetragonal system there are two: a simple P and a body-centred I. In the orthorhombic there are four: a simple P, face-centred on the top face C, body-centred I and face-centred on all faces F. In the hexagonal, rhombohedral and triclinic systems there is just one simple lattice for each, but two lattices are possible in the monoclinic system—a simple lattice P and one which is centred on the end faces, C.

CHAPTER II

Experimental Demonstration of Crystal Diffraction

WE HAVE shown in the previous chapter how studies of the external form of crystals led to the comprehensive examination of crystal symmetry and its classification in terms of the thirty-two crystal classes. Before proceeding to extend these ideas about symmetry to the case of continuous extended arrangements in three dimensions, we must examine the direct experimental evidence that solids are in fact built up in this regular fashion. The evidence arose from a suggestion by von Laue in 1912 which was designed to demonstrate that X-rays were a type of wave motion.

It had been realized some years earlier that with a knowledge of Avagadro's number it was possible to calculate from the density of a metal or salt the volume of space occupied by a single atom or molecule. For example, the density of metallic iron is 7·9 so that a gram atomic weight, 55·8 g, would have a volume of 7·06 cm^3. This will be the volume occupied by 60×10^{22} atoms, so that the volume per atom is $11·7 \times 10^{-24}$ cm^3. If we suppose, as the simplest possibility, that each atom occupies a unit cube, then the dimension of this cube would have to be about $2·2 \times 10^{-8}$ cm, i.e. 2·2 Å. It was von Laue's belief, based on the rather inconclusive results of early experimenters who had tried to demonstrate that X-rays could be diffracted by a straight slit, that X-rays might have a wavelength of this order of magnitude. He concluded therefore that they would be diffracted by the regular arrangement of atoms in a crystal in rather the same way that visible light (of wavelength about 6000 Å) is diffracted by the rulings on an ordinary diffraction grating. The main difference would be that the latter has only a one-dimensional periodicity whereas the crystal would have a regular arrangement in three-dimensions. The geometrical conditions for diffraction would therefore be more stringent.

Laue suggested to Friedrich and Knipping that they should investigate whether a crystal would diffract X-rays, and their experiments using a crystal of copper sulphate in the first instance were completely successful. An account of Laue's reasoning which led to the experiment, and of the experiment itself, appears in Paper 1.

The experiment demonstrated quite conclusively that the diffracted X-rays were a form of wave motion and that the crystals had a periodic structure in three dimensions. However, the investigators were not at that time convinced of the relationship between the incident and diffracted rays and for a long time held on to a notion that the latter rays consisted of fluorescent radiation which was "characteristic" of the crystal. Indeed they made their original selection of crystals specifically to include heavy elements, which were known from Barkla's work to give good fluorescence. It was the work of W. H. and W. L. Bragg, father and son, which led to a clear interpretation of the German experiments. This becomes clear in W. L. Bragg's paper to the Cambridge Philosophical Society entitled "The Diffraction of Short Electromagnetic Waves by a Crystal" which appears as Paper 2. It was the work described in this paper which really led to the development of X-ray diffraction as a means of studying the internal structure of crystals, i.e. finding the size and shape of the unit of pattern and the details of the arrangement of atoms within this unit.

The reader who is interested in the historical circumstances of Laue's work is referred to a first-hand account of these events by P. P. Ewald (1962) in *Fifty Years of X-ray Diffraction*. At that time Ewald was, like Laue, working in Sommerfeld's Institute for Theoretical Physics in Munich and was completing his doctoral thesis on "The optical properties of an anisotropic arrangement of isotropic resonators". Ewald was concerned with the behaviour of visible light but was able to interest Laue in the relevance of this work to the latter's own problem of the passage of much shorter wavelengths (which was what he believed X-rays to be) through crystals. In due course, after the successful X-ray experiment, Ewald developed his own way of interpreting the results for short waves in terms of the "reciprocal lattice", and "sphere of reflection" which are always associated with his name. We shall refer to this again in Chapter IV.

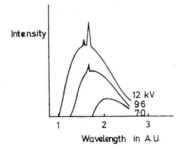

FIG. 8. The X-ray spectrum from a nickel target as a function of the applied voltage. A continuous spectrum is produced together with peaks, characteristic of the target, which appear above a certain threshold voltage.

The crucial point which becomes clear in W. L. Bragg's paper (2) is the realization that the X-rays which were incident on Laue's crystal had a continuous spectrum of wavelengths, i.e. a broad distribution of the kind which is illustrated in Fig. 8, and that the action of the crystal was simply to "pass on" just those wavelengths for which the scattered contributions from individual atoms in the regular crystal array were in phase. At the same time Bragg presented an alternative, but entirely equivalent, way of looking at the diffraction process itself. He visualized the process as a reflection of the incident beam by the internal planes of atoms in the crystal. The condition which had to be satisfied was expressed in what has since come to be known as the Bragg Law

$$n\lambda = 2d \sin \theta$$

where, as indicated in Fig. 9, d is the distance of separation between successive planes in the crystal.

The reader will notice that the law is not stated exactly in this form in Paper 2, but as the less familiar $n\lambda = 2d \cos \theta$, which

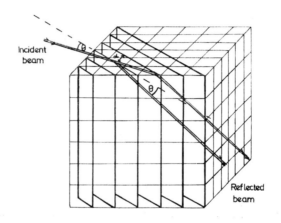

Fig. 9. The build-up of a diffracted beam by reinforcement of the contributions reflected from successive crystal planes: d is the interplanar spacing and θ the glancing (Bragg) angle.

arises because Bragg used θ to denote the angle of *incidence*, instead of the current practice of using the *glancing* angle, or, as it has now come to be known, the "Bragg" angle. The law indeed is a statement of the fact that a "reflected" beam will be produced, in a direction which is inclined to the plane-normal at the same angle as the incident beam, if the contributions scattered from successive planes are in phase. We shall show later that this is exactly the same condition as is implied in the three Laue equations, which express the conditions that equivalent *points* in all unit cells shall give in-phase contributions to the scattered beam. Another quite important problem which was cleared up in Bragg's paper relates to the structure of zinc blende. Laue had not been able to interpret the pattern of scattered X-rays from this crystal in detail, and had found that certain of the expected diffraction spots were not present. Bragg was able to show that

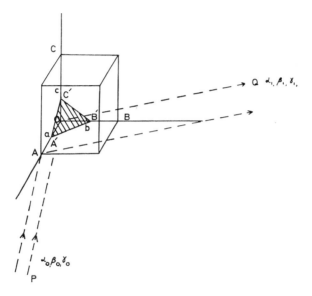

Fig. 10. Construction to demonstrate the equivalence of the Laue and Bragg conditions for the formation of a diffracted beam. The path from P to Q via any point on the shaded plane A'B'C' is one wavelength shorter than that from P to Q via the origin O.

this was because the structure of zinc blende was based on a face-centred lattice, with four molecules per unit cell, and not on a simple cube which would contain only a single molecule. Bragg's conclusion accounted exactly for the occurrence of all the observed spots and it can be said that this consideration of the zinc blende problem was really the first step in using Laue's discovery as a tool for studying the structure of crystals.

The equivalence of Bragg's Law and the set of three Laue equations can be seen by considering Fig. 10 for the case of a monochromatic beam of X-rays falling on a crystal.

Here the point O is the origin of a unit cell in the crystal structure and A, B, C are equivalent points along the three axes OX, OY, OZ of the unit cell, so that $OA = a$, $OB = b$, $OC = c$ where a, b, c are the unit cell edges. Consider that a plane wave of monochromatic X-rays from a distant point P is falling on the assembly of points and let us deduce the conditions which are necessary in order to give a diffracted wave in the direction of Q. Let the direction cosines of OP, OQ be α_0, β_0, γ_0 and α_1, β_1, γ_1 respectively.

The scattered rays from the points O, A will be in phase if

$$a(\alpha_1 - \alpha_0) = h\lambda \tag{2.1}$$

where λ is the X-ray wavelength and h is an integer. Similarly the contributions from O, B will be in phase if

$$b(\beta_1 - \beta_0) = k\lambda \tag{2.2}$$

and those from O, C if

$$c(\gamma_1 - \gamma_0) = l\lambda. \tag{2.3}$$

These three equations are termed the Laue equations and together constitute the conditions which have to be satisfied if a diffracted beam is to appear in the direction of Q. In general, that is for any arbitrary set of incident direction cosines, all three equations will not be satisfied, since any direction of diffraction is completely specified by *two* of its direction cosines. This means that only for certain specified directions of incidence will any diffracted beam be formed when a monochromatic beam of incident X-rays is used.

From equations (1), (2), and (3) we have by squaring and adding

$$2 - 2(\alpha_0\alpha_1 + \beta_0\beta_1 + \gamma_0\gamma_1) = \left(\frac{h^2}{a^2} + \frac{k^2}{b^2} + \frac{l^2}{c^2}\right)\lambda^2 \tag{2.4}$$

since for any set of directions cosines $\alpha^2 + \beta^2 + \gamma^2 = 1$. By a well-known theorem, we know that if 2θ is the angle between the directions OP, OQ then

$$\cos 2\theta = \alpha_0\alpha_1 + \beta_0\beta_1 + \gamma_0\gamma_1.$$

Thus equation (4) becomes

$$2 - 2\cos 2\theta = 4\sin^2\theta = \left(\frac{h^2}{a^2} + \frac{k^2}{b^2} + \frac{l^2}{c^2}\right)\lambda^2$$

and

$$\lambda = \frac{2}{\sqrt{\dfrac{h^2}{a^2} + \dfrac{k^2}{b^2} + \dfrac{l^2}{c^2}}}\sin\theta$$

It can easily be shown† that

$$\left(\frac{h^2}{a^2} + \frac{k^2}{b^2} + \frac{l^2}{c^2}\right)^{-\frac{1}{2}}$$

is the interplanar spacing d of the set of planes whose "Miller indices" are hkl, i.e. the separation between the plane which makes intercepts a/h, b/k, c/l on the three axes and a parallel plane through the origin. Hence we have shown that the Laue equations are equivalent to $\lambda = 2d\sin\theta$, and our proposition will be proved if we can show that it is indeed the crystallographic hkl plane which is symmetrically inclined, at angle θ, to the directions OP, OQ in Fig. 10.

In order to demonstrate that this is so, let us mark off points A', B' and C' such that $OA' = a/h$, $OB' = b/k$ and $OC' = c/l$; the crystallographic indices of plane $A'B'C'$ will therefore be

---

† Let the perpendicular from the origin onto the hkl plane be of length d and have direction cosines α, β, γ. By considering the planar sections which contain this perpendicular and, in turn, each of the axes it can be seen that $d = (a/h)\alpha = (b/k)\beta = (c/l)\gamma$. But for any set of directions cosines $\alpha^2 + \beta^2 + \gamma^2 = 1$, so that $(h^2/a^2)\,d^2 + (k^2/b^2)\,d^2 + (l^2/c^2)\,d^2 = 1$, thus leading to the value of d quoted above.

hkl. In the figure we have chosen the values $h = 2$, $k = 2$ and $l = 3$ so that $A'B'C'$ is the 223 plane. Equation (1) expresses the fact that the path PAQ is h wave-lengths shorter than the path POQ. It follows therefore that the path $PA'Q$ will be *one* wavelength shorter than POQ and, in just the same way, it can be seen that paths $PB'Q$, $PC'Q$ are also *one* wavelength shorter than POQ. This means that the pathlengths from P to Q via A', B' and C' are all *equal*, which is the same as saying that the direction OQ is the reflection in the plane $A'B'C'$ of the direction OP and that the bisector of the angle POQ is normal to this plane. We have therefore established the equivalence of the Laue and Bragg conditions.

CHAPTER III

Internal Crystal Symmetry and Space Groups

IN OUR discussion in Chapter I of the symmetry of crystal faces, or of the groups of points on a stereogram which can represent these faces, we regarded a symmetry element as something which after "operating" on a set of points produced a set which was identical in all respects to the original set. In Chapter II we presented the first direct experimental evidence that solids have a regular extended structure in three-dimensions. It becomes plausible to assume therefore that the fourteen Bravais lattices, which describe the possible ways of assembling *infinite* arrays of identical points, will provide the basis for a method of classifying the structures of solids. In attempting to catalogue all the possible types of atomic arrangement we can begin by taking each of the cubic lattices, extended continuously in three-dimensions, and then placing at each lattice point an assembly of points which has the symmetry of a chosen cubic point group. This will produce a particular variety of extended structure: the dimensions and the choice of atoms which constitute the point unit can, of course, be varied at will. We do this in turn for each of the cubic point groups and then continue by combining each tetragonal point group, in turn, with each of the two tetragonal Bravais lattices, and so on. If we collect together all the resulting combinations we shall have succeeded, with some reservations, in producing a list of all possible varieties of extended arrangement in three-dimensions. However, further consideration shows that our list will be incomplete because we shall find that we have to widen our conception of symmetry elements as soon as we begin to consider extended arrangements.

As a simple example of what is involved consider in Fig. 11 (i) a tetragonal grouping of points P, Q, R, S placed about

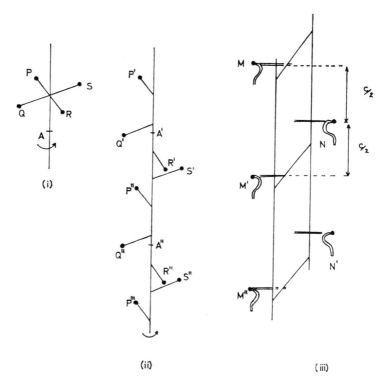

FIG. 11. Diagrammatic representation of (i) a fourfold axis of rotation, (ii) a fourfold screw axis, (iii) a glide plane.

point A through which the vertical axis passes in the figure. This means that for every point P, for example, there is a corresponding point Q, related to P by a 90° rotation about the vertical axis. Now consider the arrangements of points P'. Q', R', S' which are grouped about point A' in Fig. 11 (ii). This is part of an extended distribution of points and there is an exactly similar fourfold grouping P'', Q'', R'' and S'' associated with point A'' further down the vertical axis. This group of points does not possess the fourfold axis which was present in (i). However, we find that this arrangement of points possesses a new kind of symmetry axis, namely a vertical *screw* axis whose function is to rotate through 90° followed by translation through one quarter of the distance of separation $A'A''$. This procedure moves P' to Q' and then, successively, to R', S' and P''. Such an axis becomes possible with the extended distribution of points because S' (for example) is able to be transposed into a point *in the next set*. A screw axis would be meaningless with a single set of points P', Q', R' and S' since, because of the translation which is involved in addition to the rotation, its operation does not reproduce the same four points but involves those in successive sets.

In an exactly similar way we can, in an extended distribution, have points which are correlated by a so-called *glide* plane, as in Fig. 11 (iii). In this case the grouping represented at M is transposed into that at N by a reflection across the intervening shaded plane followed by a downward translation through a distance $c/2$. The same combined operation of reflection and translation then, in turn, operates on N to produce the grouping at M' which is identical with that at M but is one unit c lower down, in the next cell.

In order to produce all possible extended arrangements of points in space, or "space groups" as they are called, our procedure is therefore as follows. With each Bravais lattice we use, in turn, each appropriate point group, first in its ordinary form and then by replacing any axes and reflection planes by *screw* axes and *glide* planes. In some cases two different combinations of Bravais lattice and point group will produce the same, or an equivalent structure, so that the total number of different "space groups" is not so large as we might at first think. In fact there are 230 of them, and they were first identified by symbols proposed by Schoenfliess. These symbols were replaced by an alternative notation due to Hermann and Manguin which is now accepted internationally and we shall use this notation in our subsequent discussion. A symbol, for example $P2_1/m$, consists of two parts. The initial capital letter identifies the type of Bravais lattice—thus P represents a primitive cell, C is a cell which is face-centred on the c face and I represents a cell which is body-centred. This

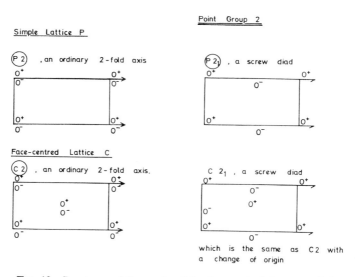

FIG. 12. Summary of the mode of development of the monoclinic space groups for the point groups 2 and m. Point group 2 gives rise to $P2$, $P2_1$, and $C2$: point group m gives Pm, Pa, Cm, and Cc (see facing page).

capital letter is followed by one, two, or three symbols which describe the elements of symmetry which lie along special directions in the crystal. Thus the symbol 2_1 in our example $P2_1/m$ means a twofold screw axis in the direction of the unique axis of the monoclinic system and the symbol m preceded by the stroke / means an "ordinary" reflection plane which is perpendicular to this unique axis.

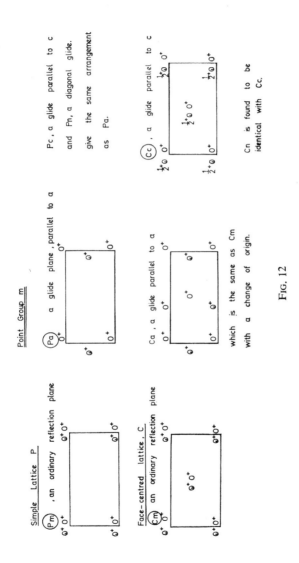

Simple Lattice P

Pm, an ordinary reflection plane

Face-centred lattice, C

Cm, an ordinary reflection plane

Point Group m

Pa, a glide plane, parallel to a

Ca, a glide parallel to a

which is the same as Cm
with a change of origin.

Pc, a glide parallel to c
and Pn, a diagonal glide.
give the same arrangement
as Pa.

Cc, a glide parallel to c

Cn is found to be
identical with Cc.

Fig. 12

The crystal class or point group to which any particular space group gives rise is obtained by omitting the lattice symbol and replacing all the screw axes by the corresponding simple rotation axes and all the glide planes by simple reflection planes. Thus our example $P2_1/m$ and also, as another example, $P2/c$ both become crystal class $2/m$. Crystals whose internal periodic structure corresponded to either of these space groups would show the external point group symmetry of $2/m$ in, for example, the arrangement of their faces and their physical properties. For further details of this symbolism the reader is referred to textbooks of crystallography (e.g. Phillips, 1963). Here, we shall content ourselves with giving an example, by taking the special case of the monoclinic system, in order to make it clearer exactly how the total number of space groups is deduced.

In the monoclinic system there are three point groups as indicated by the stereograms in Fig. 5. They are designated respectively as: 2, which indicates a diad axis as the sole symmetry element; m, which is simply a reflection plane; and $2/m$, which comprises a diad axis which is perpendicular to a reflection plane. As we have seen from Fig. 7, there are two Bravais lattices in the monoclinic system, namely a simple lattice P and one, C, which is centred on the end faces. How many space groups can we produce? The various possibilities for point groups 2 and m are summarized in the schematic projections drawn in Fig. 12. We see in fact that point group 2 gives rise to the three space groups $P2$, $P2_1$ and $C2$: point group m produces the four space groups Pm, Pa, Cm and Cc. By a similar procedure we should find that point group $2/m$ gives the six groups $P2/m$, $P2_1/m$, $P2/a$, $P2_1/a$, $C2/m$ and $C2/c$. Accordingly therefore the total number of monoclinic space groups is $3+4+6 = 13$.

Proceeding in this way it has been established that the total number of possible space groups included in all the crystal systems is 230. This number of space groups is distributed amongst the seven crystal systems according to Table I which also lists the number of point groups and Bravais lattices in each system.

We shall see later that the problem of finding out to which

TABLE I. DISTRIBUTION OF POINT GROUPS, BRAVAIS LATTICES AND
SPACE GROUPS AMONG THE SEVEN CRYSTAL SYSTEMS

Crystal system	Number of point groups	Number of Bravais lattices	Number of space groups
Triclinic	2	1	2
Monoclinic	3	2	13
Orthorhombic	3	4	59
Trigonal	5	1	25
Hexagonal	7	1	27
Tetragonal	7	2	68
Cubic	5	3	36
Total	32	14	230

particular space group the structure of a given crystal belongs is
one which can be solved directly by X-ray diffraction techniques.
Our study of space groups is therefore the foundation for the
first stage in the process of determining the structure of an
unknown crystal.

CHAPTER IV

The Principles of Crystal-Structure Determination with X-rays

AN OPTICAL diffraction grating provides a simple one-dimensional analogy which we can use as a basis for developing an understanding of the way in which the regular three-dimensional structure of crystals can be determined by using X-rays. It is well known from ordinary optical theory (see, for example, Jenkins and White, 1957) that the spectra from a diffraction grating, illuminated by parallel monochromatic light, occur at angular *positions* which are dependent on the grating "period", i.e. the distance between the centres of neighbouring lines. On the other hand the distribution of *intensity* among the different orders of spectra is determined by the width and transmission function of the individual lines, which will depend on the nature of the instrument and procedure used for ruling the lines. If, as a very simple example, we take the case of a grating for which the grating period is n times as large as the width of an individual line then we find that the orders of spectra n, $2n$, $3n$, . . . are all absent. In detail it can be shown quite generally (e.g. Stone, 1963) for any grating that the distribution of energy among the different orders of spectra is the Fourier transform of the transmission factor of the grating line. For example, a grating which gives a transmission of light varying with distance in a simple harmonic manner will yield only first-order spectra, one on each side of the direction of the incident beam: such a grating will not transmit any zero-order light.

The same principles of diffraction apply to our three-dimensional crystal gratings but there is one particularly important difference. As we have already seen in Chapter II, there are three Laue equations which have to be satisfied in order that any individual spectrum may be produced from the crystal. This means that the experimental methods which we use for measuring the size of the unit cell and the intensities of individual spectra must provide an extra degree of freedom, namely motion of the crystal, in order that the spectra can be observed. We shall return later to the details of the way this is carried out in practice but we can state now that in principle we can examine all the three-dimensional spectra and that we can identify each of them by the three Miller crystallographic indices hkl. In general we shall find by experiment that some combinations of indices are missing, even though we may have available extremely sensitive means of detecting very weak intensities, and it can be shown that from a systematic study of such absences it is possible to determine the space group of the crystal. The absences are of two types. "General" absences, which apply to all hkl reflections, give clues to the type of Bravais lattice. For example, with a body-centred lattice the sum $h+k+l$ is always even, whereas with a lattice which is centred on all faces, h, k, l are either all odd or all even: if the lattice is only centred on the A face then the condition is that $k+l$ must be even. As well as these general absences, or "extinctions" as they are often called, there will be "special" absences which refer only to particular groups of reflections. Thus, if the space group includes a twofold screw axis parallel to the x-axis then the only $h00$ reflections which occur will be those for which h is even. In a similar way it can be deduced that if there is a glide reflection plane normal to the x-axis and including a translation $b/2$ in the y-direction, then the only $0kl$ reflections which can occur are those for which k is an even number. By examining the list of observed spectra to discover extinctions of both "general" and "special" kinds we are able to derive the space group of the crystal and this is an essential preliminary step towards determining its detailed structure. A very simple example of this procedure is described in Paper 3 by W. L. Bragg, which was published in 1914 and demonstrates that in ordinary metallic copper the atoms are arranged on a face-centred lattice and not on a simple cubic lattice. At this early date the convention of describing the observed reflections in terms of their Miller indices was not fully established and the author speaks of first, second and third order reflections from particular crystal planes. The reflection which he regards as the first-order reflection from the 100 planes because it is the lowest order which *occurs*, we should now describe as (200). Bearing this in mind the reader will be able to correlate these early results with the foregoing systematic discussion. In particular, he should be able to conclude that in Fig. 2 of Bragg's paper the indices of the reflections, in ascending order of their θ values, are (111), (200), (220), (222), . . . , etc. The reflections which are designated as (100) and (110) when this standard nomenclature is used are absent in the case of copper because the atoms lie on a face-centred lattice.

Let us now return to an examination of the experimental methods of observing, and measuring, the diffraction spectra. In the study of copper which we have just discussed, the crystal, which had faces measuring about 1 cm each way, was mounted with one of its faces lying on the axis of an X-ray spectrometer, such as was described by Bragg (1914b) in an earlier paper. It was then turned into the appropriate position θ for a particular spectrum to be produced and the reflected X-rays were detected in the ionization chamber, assuming that this had been placed in the correct angular position 2θ to receive the reflection. This successive positioning of crystal and detector is a relatively slow procedure, but it is easy and straightforward for large crystals and particularly for those which have a simple structure. However, it was not long before other, more flexible, methods were developed. In 1916, rather by chance while they were searching for evidence of spatial structure in the outer electrons of the atom, Debye and Scherrer (1916) discovered the so-called "powder method" which has proved so valuable, especially in the

industrial applications of X-ray diffraction. In order to understand the principle of this method let us consider the illustration
in Fig. 13 (i) which shows a large crystal oriented to give a particular reflection on the X-ray ionization spectrometer. The incident
X-rays travel along *OP* and are reflected by planes having a

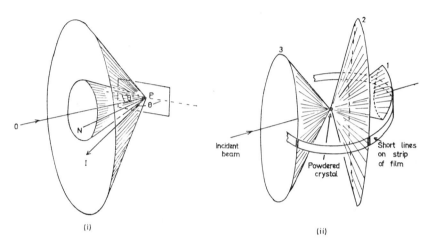

FIG. 13. The basis of the "powder method" of X-ray diffraction.
In (i), for a given crystal plane, a crystal fragment whose normal
PN lies on the surface of the inner cone, with its axis in the incident
beam direction *OP*, will give a diffracted beam such as *PI*, lying on
the surface of the outer cone. In (ii) the diffracted conical haloes
1, 2, and 3 from a powdered crystal, corresponding to three
different sets of crystal planes, intersect an equatorial film strip in
short "diffraction lines".

normal *PN* to give a diffracted beam in the direction *PI*. The two
angles shown will be equal to θ and the Bragg equation has to be
satisfied. If we now imagine the crystal to rotate around the
direction *OP* in such a way that the normal *PN* to the reflecting
plane remains at a constant angle θ to *OP*, then reflection of the
X-ray beam will continue to take place. However, the reflected
beam will not remain in the direction *PI* but will be found to

move around the surface of a cone which we generate by rotating *PI* about *OP*; at the same time the planar normal *PN* generates the surface of the cone which has half the semi-angle of the former one. It follows therefore that if instead of a single-crystal we have a random distribution of crystal fragments, such as exist in a powder, then all the fragments whose normals (for a given plane) happen to lie along the inner cone in Fig. 13 (i) will give a reflection and this reflection will lie somewhere around the surface of the outer cone. When we consider all possible crystal reflections it is clear that if we allow an X-ray beam to fall on a small sample of powdered material then each reflecting plane will produce its own diffracted "halo" as in Fig. 13 (ii). If we place a narrow strip of photographic film around the sample then its intersection with these conical halos will produce short curved lines on the film, as the figure indicates. The curvature will, of course, vary with the distance around this circumferential film. This is the well-known "powder photograph" or Debye–Scherrer pattern. The same method of studying diffraction from a powder was discovered independently about a year later by Hull in the U.S.A. and the latter's account of the method, published in 1917, appears as Paper 4. As the author of this paper points out, the original Bragg method with the ionization spectrometer was very simple when single crystals of sufficient perfection were available. Failing this, however, the next-best practical alternative is the perfect chaos of planes which exists in a crystal powder, where all the crystal planes have an equal opportunity of reflecting.

It will be realized, of course, that the simplicity of the powder method is only achieved at the expense of losing directional information. The only parameter of which the powder method takes account is interplanar spacing. Thus if in a crystal structure there are two sets of planes in quite different directions but which have the same separation, then the powder method will not be able to distinguish between them. Exact coincidences occur in cubic crystals where, for example, the reflections 322 and 410 are identical in position, and with structures of lower symmetry there will always be random overlapping between sets of planes whose spacings are closely similar. This restriction is a great drawback when the powder method has to be employed with crystals of low symmetry, since these generally produce a very large number of closely spaced lines on the diffraction photograph. Many of these lines cannot be separately identified, quite apart from the impossibility of measuring their individual intensities.

In spite of the great usefulness of the powder photograph for identifying phases, studying the changes of unit cell size with temperature, determining the actual sizes of crystallites and for many other applications, it is almost always essential to examine single crystals when the structure of a new substance is being examined. For over thirty years this examination was done almost entirely by photographic methods based on what is known as the "rotating crystal" method, or simply as the "rotation photograph". Indeed in spite of the increasing use of radiation counters as detectors, the rotation photograph is very largely used, particularly for all preliminary and exploratory work, even today. One of the first detailed accounts of this method, and certainly the first full account in English, was published by Bernal in 1926 and the first part of this account is reproduced as Paper 5.

This paper utilizes the concept of the reciprocal lattice, which was first discussed by Ewald as we have already mentioned, in order to determine quantitatively the conditions for production of a reflected beam from the rotating crystal. The idea of reciprocal space arises from a need to have a convenient way of representing a series of parallel planes, bearing in mind that the reflection of an X-ray beam by such planes is determined by both their spacing and their orientation. In the reciprocal lattice we represent each family of planes by a point which is placed at a distance from the origin equal to the reciprocal of the interplanar spacing and in a direction which is normal to the planes. Each set of planes *hkl* is represented by its own point and the whole assembly of points constitutes a reciprocal lattice. It is the fact that the *reciprocal* of the spacing is employed that gives the valuable geometrical relationships in connection with the diffraction pattern of a crystal which Bernal describes in his paper. As the paper shows, a reflection corresponding to a particular point on the reciprocal lattice occurs as this point passes through the "sphere of reflection", a sphere whose surface goes through the origin of the reciprocal lattice and has the direction of the in-

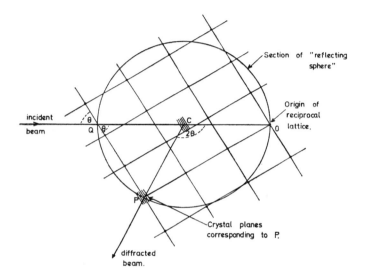

Fig. 14. Reciprocal lattice and reflection sphere. The crystal is rotating about a vertical axis through O, the origin of the reciprocal lattice, and reciprocal point P is just passing through the reflecting circle. The incident X-rays make a glancing angle θ to the crystal planes represented by P such that $\sin \theta = OP/OQ = 1/d \div 2/\lambda$ by definition of the reciprocal lattice. Hence $\sin \theta = \lambda/2d$ so that reflection takes place according to the Bragg Law, in the direction CP.

cident X-ray beam as a diameter. For simplicity we can examine the geometrical arrangement in two dimensions and it is indicated in the equatorial plane by Fig. 14 in which the reciprocal net is considered to rotate across the circle defined by the incident X-ray beam. If the crystal is rotated about a vertical axis then individual planes will come successively into reflecting positions.

Thus the point *P* in the figure lies on the circle and will give a reflection. An example of the resulting picture which is produced on a cylindrical film placed around the crystal is shown in Fig. 15. If the vertical axis of rotation is the *c*-axis of the crystal then the

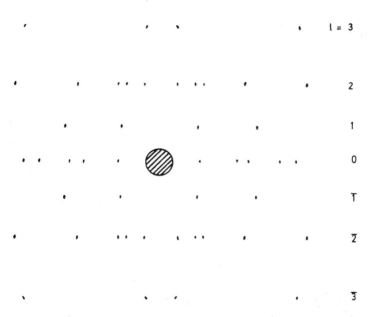

FIG. 15. The arrangement of diffraction spots on a typical rotating crystal diffraction photograph. The spots lie on horizontal layer lines, each associated with a horizontal layer of points in the reciprocal lattice. These points pass, in turn, through the sphere of reflection as the crystal rotates about a vertical axis. The large diffuse central spot is caused by the undeviated beam.

successive horizontal "layer lines" in such a picture will have *l* indices of ...$\bar{3}$, $\bar{2}$, $\bar{1}$, 0, 1, 2, 3, ... and can be indexed as such. Each line will include all the points in the corresponding layer of the reciprocal lattice. The second feature of importance is the layout on the film of those points situated on any vertical row in the reciprocal lattice, i.e. points which have constant values of h, k. The reflections corresponding to these points will be distributed on the film as shown by the curves in Fig. 16: this figure indicates how the curves for h, k constant superpose on the horizontal layer lines of constant l. From a consideration of features of this kind it becomes possible to "index" the spots on the rotation photograph, i.e. to ascribe the correct indices h, k, l to the individual reflections, thus making it possible not only to

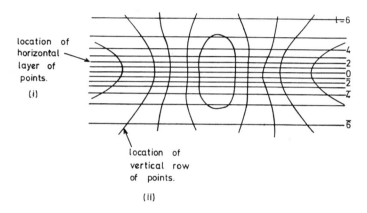

FIG. 16. The layout on a cylindrical film of (i) the horizontal layer lines corresponding to horizontal layers of points in the reciprocal lattice and (ii) the curves of variable shape which correspond to vertical rows of points, i.e. points with the same values of h, k indices but different values of l.

state which particular hkl spectra are missing but also to determine the relative intensities of those which are present. In practice other more specialized techniques, particularly moving-film methods, are also used to improve this process of collecting intensity data. When photographic methods of any kind are used it is, of course, necessary to measure the intensity of blackening of the photographic film by means of a photometer if precise measurements of the intensities are to be made. However, for the first twenty or thirty years of the development of these methods practically all intensities were recorded simply by visual estimates, such as very weak, weak, average, . . . , strong, etc. Nowadays, as techniques have improved and the order of accuracy aimed at has increased by several orders of magnitude, radiation detectors such as Geiger–Muller counters are often employed in order to provide very accurate direct measurements of the intensities of the diffracted X-ray beams. It must be realized, though, that a counter detector is in some ways a less convenient and less flexible device than a photographic film. In particular, the counter has to be rotated in three dimensions about the pole of the spectrometer in order to collect in turn the separate diffracted beams, whereas a single cylindrical film can collect all the beams within a very large angular range. Consequently, the use of counters has encouraged and necessitated the development of automatic diffractometers in which the movements of both crystal and counter are programmed in advance, by means of devices such as punched paper tapes.

We conclude this chapter by recalling that the aim of our investigations so far has been to measure the positions and intensities of all the diffraction spectra. The angular positions of the spectra will have enabled us to deduce the dimensions of the unit cell of the crystal structure and an examination of the indices of any missing spectra will have allowed us to deduce the space group of the structure. The next stage in our analysis is to use the quantitative measurements of intensity to determine how atoms are arranged within the unit cell, that is to locate in three-dimensional space the individual atoms and molecules which the structural unit contains. We shall examine the ways by which this can be attempted in the next chapter.

CHAPTER V

Fourier Analysis of Crystal Structures

WE HAVE already noted that for an ordinary optical diffraction grating the spectral pattern is the Fourier transform of the transmission factor of the lines which make up the grating. We shall now consider in more detail the corresponding relationship for a three-dimensional grating.

For the scattering of X-rays by a single *electron* it is well known from the work of J. J. Thomson (see, for example, Compton and Allison, 1935) that the amplitude E of the scattered wave at distance r is

$$E = E_0 \frac{e^2}{mc^2} \frac{1}{r} \sin \chi \qquad (5.1)$$

where E_0 is the amplitude of the incident plane wave and χ is the angle between the direction of scattering and the direction of the electric field vibration in the incident wave. When we consider the scattering by a single *atom* we can say that the scattering is greater by a factor f than that from a single electron and we call this factor the "atomic scattering factor". Since the electrons in an atom are distributed over a volume whose linear dimensions are comparable with X-ray wavelengths, equal to about 1·5 Å, it follows that f will vary with the angle of scattering 2θ. As θ increases f decreases, because of the increasing phase-differences which arise for the scattered contributions from different parts of the atom. When we consider the resultant scattered contribution from a *unit cell* of a crystal structure it is the phase differences between neighbouring atoms which become of predominant importance, because these are much larger than those between the different parts of an individual atom. Taking these into account we obtain for F_{hkl}, the structure amplitude factor of the unit cell

$$F_{hkl} = \sum_n f_n \exp 2\pi i(hx/a + ky/b + lz/c) \tag{5.2}$$

where the summation is made over the n atoms in the unit cell. In more general terms we can consider the unit cell to have a continuous distribution of electron density, instead of discrete individual atoms, and if we write this as ρ_{xyz} at the point x,y,z then it follows that

$$F_{hkl} = \int_0^a \int_0^b \int_0^c \rho_{xyz} \exp 2\pi i(hx/a + ky/b + lz/c).\,dx\,dy\,dz \tag{5.3}$$

The function ρ_{xyz} is, of course, a periodic function, with spatial periods of a, b, c respectively in the three crystallographic directions, and it can therefore be expressed as the sum of a three-dimensional series of simple harmonic terms in accordance with Fourier's Theorem. Thus

$$\rho_{xyz} = \sum_{h'}\sum_{k'}\sum_{l'} C_{h'k'l'} \exp 2\pi i(h'x/a + k'y/b + l'z/c) \tag{5.4}$$

where $C_{h'k'l'}$ is the coefficient of the Fourier term which has spatial periods of a/h', b/k', c/l'. The diagrams in Fig. 17 illustrate the contribution which is made to the electron density over the unit cell by two particular Fourier terms, namely (200) and (320).

If we now combine equations (5.3) and (5.4) we have

$$F_{hkl} = \int_0^a \int_0^b \int_0^c \sum_{h'}\sum_{k'}\sum_{l'} C_{h'k'l'} \exp 2\pi i(h'x/a + k'y/b + l'z/c) \times$$
$$\times \exp 2\pi i(hx/a + ky/b + lz/c)dx\,dy\,dz \tag{5.5}$$

When the integration is performed the product of each pair of exponential terms yields zero except for the particular case when $h' = -h$, $k' = -k$, $l' = -l$. In this special instance the value of F_{hkl} simply reduces to the product of *abc*, the unit cell volume, by the Fourier coefficient $C_{\bar{h}\bar{k}\bar{l}}$. Bearing in mind that *hkl* and $\bar{h}\bar{k}\bar{l}$ represent the same direction we arrive at the conclusion that each Fourier coefficient C_{hkl} in the expression for the electron density in (5.4) is proportional to the value of the structure amplitude

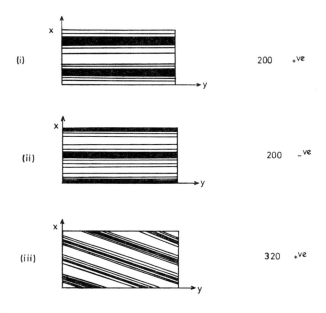

(i)   200   +ve

(ii)   200   -ve

(iii)   320   +ve

FIG. 17. Representation of the electron density in a unit cell, shown in two dimensions, as a series of Fourier terms. The diagrams in the figure show (i) the contribution from a 200 term with positive amplitude, (ii) contribution from a negative 200 term, (iii) contribution from a positive 320 term.

factor F_{hkl}. If, therefore, from our experimental values of the intensities of reflections we can find the value of each F_{hkl} we shall be able to deduce the value of ρ at any point *x, y, z* by means of equation (5.4).

Before we can implement this conclusion, and thus discover the way in which ρ varies over the unit cell, we have to overcome two difficulties. First of all we have to emphasize that F_{hkl} is a complex quantity, having amplitude and phase. The experimental diffraction intensities are proportional to FF^* or $|F|^2$ and by making due allowance for geometrical factors we can indeed find the value of $|F|^2$ for each reflection. However, this means that although we can find the *magnitude* of each F_{hkl} we do not know its phase and there is indeed no direct way of finding this phase by experiment: this is a fundamental limitation of our method. For structures which possess a centre of symmetry the problem is less general and the phase of F can be only 0° or 180° relative to the phase of the contribution scattered by an atom placed at the origin of co-ordinates so that our difficulty reduces to that of ascribing either a + or − sign to each of the numerical values of F which we have deduced from the intensities of the spectra. We shall return shortly to consider some of the indirect methods whereby we may find out these individual signs. Our second comment is that, even if the signs of the F values are established, there remains a substantial computational problem. This second difficulty was a very restrictive one in the early days of structural analysis but has largely vanished now that electronic computers are widely available.

In a typical structure determination about a thousand reflections may be measured so that in order to determine the electron density at a *single point* x, y, z in the unit cell, by using equation (5.4), we shall have to add together a thousand terms. In the early days of X-ray diffraction analysis the investigation was almost invariably restricted to a study in two dimensions, leading to knowledge of a projection of the electron density on some chosen plane in the unit cell. For example, we can produce a projection of the electron density on the 001 plane by simply making measurements of the hk0 reflections, that is the reflections which appear in the equatorial plane if we mount a crystal with its c-axis vertical and then rotate it about the vertical axis. To produce such a projection might need two hundred reflections and even this results in a great deal of numerical computation if we are to obtain a detailed picture of the projected electron density over a closely space mesh of points x, y in the plane of projection. It is not surprising therefore that great effort and ingenuity was devoted by early workers in order to minimize the tedious labour, and time, which was taken up in producing a picture of electron density in the crystal even when the signs of the structure amplitude factors F had, somehow, been deduced.

An account of an optical procedure for performing the addition of the Fourier terms which was used by W. L. Bragg in 1929 appears as Paper 6. The method can be understood from Fig. 17, from which it is developed very simply. For each set of indices, such as 200 and 320 which are illustrated in the figure, a photographic plate is prepared with dark and light bands of appropriate spacing, by photographing a row of opaque rods placed at a distance apart equal to twice their diameters. By using suitable illumination it was possible to produce a plate whose transmission consisted of a set of simple harmonic bands superimposed on a constant distribution. By exposing a sheet of photographic paper to the light transmitted by a series of such plates in succession, and arranging that each exposure time is proportional to the value of F for the corresponding spectrum, a "microphotograph" of the crystal structure is obtained. Bragg first applied this technique to a determination of the structure of the crystal diopside $CaMg(SiO_3)_2$.

Many other devices, some of them mechanical aids and others simply numerical procedures, have been used for assisting with the computational process. Perhaps the most widely used was a method which reduces the computations for a two-dimensional projection to the addition of a number of one-dimensional cosine and sine terms. The numbers are presented on cardboard strips, known after the originators of this method as Beevers–Lipson strips. The original description of this work appears as Paper 7. An individual cardboard strip is provided for each index h or k (e.g. 3) and for each value of F_{hk} (e.g. 45) and it lists, for this particular example, the values of $45 \cos 3u$ for values of u at $6°$ intervals from $0°$ to $90°$ as indicated below. The x or y co-ordinate in the unit cell which is represented by a given value of u will be equal to $u/360$, expressed as a fraction of the unit-cell edge a or b.

45 C 3 45 43 36 26 14 0 $\overline{14}$ $\overline{26}$ $\overline{36}$ $\overline{43}$ $\overline{45}$ $\overline{43}$ $\overline{36}$ $\overline{26}$ $\overline{14}$ 0

A corresponding strip carries values of $45 \sin 3u$ at equal intervals. Later on sets of strips were printed which carried values at intervals of 3° instead of 6° in order to improve the resolution of the picture of electron density which this method produces. The synthesis reduces to selecting sets of strips and adding up successive columns of figures. As an example we show in Table II a summarized table of values of F_{hk}, and then in Table III the first set of summations, where for each value of y at 3° intervals we have carried out a summation $\sum\limits_{0}^{\infty} F_{hk} \cos ky$ over all values of k for each value of h. In the particular example which we have chosen the space group permits only reflections for which h is even. We then, in turn for each value of y, carry out a summation over all values of h for each value of x at 3° intervals. There are in fact two summations of this kind to make, one for a "cos" term and the other for a "sin" term, and the sum of these gives the resultant projected electron density at any point in the projection of the unit cell. A sample of the result obtainable in this way appears in Fig. 18. This is a projection of the scattering density for resorcinol $m-C_6H_4(OH)_2$ projected on the 001 plane. It is in fact produced by neutron diffraction instead of X-ray diffraction, and represents "neutron scattering density" and not electron density, but the principles and procedure for the computation are exactly the same. We have outlined the approximate shapes of some of the atoms in Fig. 18 by drawing "contour lines" at intervals of 50 units. In practice various forms of graphical plotting were used to derive the atomic co-ordinates from density plots of this kind.

Let us now return to what we emphasized as the fundamental difficulty of applying Fourier synthesis methods to structure analysis, namely our lack of knowledge of the sign, or in the more general case the *phase*, of the structure amplitude factors. In

TABLE II. EXPERIMENTAL VALUES OF F_{hk} IN ARBITRARY UNITS

	$h=$ 0	1	2	3	4	5	6	7	8	9	10
$k=0$	20	0	23	0	−13	0	11	0	9	0	9
$k=1$	0	0	0	39	0	23	0	−17	0	−32	0
$k=2$	23	0	17	0	18	0	51	0	39	0	0
$k=3$	0	−39	0	0	0	49	0	−18	0	−41	0
$k=4$	−13	0	18	0	57	0	35	0	35	0	28
$k=5$	0	−23	0	−49	0	0	0	0	0	21	0
$k=6$	11	0	51	0	35	0	−21	0	0	0	52
$k=7$	0	17	0	18	0	0	0				
$k=8$	9	0	39	0	35	0	0				
$k=9$	0	32	0	41	0	−21	0				
$k=10$	9	0	0	0	28	0	52				

TABLE III. SUMMATIONS OF $\sum\limits_{k=0}^{\infty} F_{hk} \cos (ky)$ FOR VALUES OF y AT $3°$ INTERVALS

Values of $y=$	$0°$	3	6	9	12	15	18	21	24	27	30
$h=0$	60	57	51	42	31	23	15	11	10	16	22
$h=2$	148	143	123	96	63	28	$\bar{5}$	$\bar{3}0$	$\bar{4}7$	$\bar{5}4$	$\bar{4}8$
$h=4$	160	150	122	82	34	$\bar{1}0$	$\bar{4}7$	$\bar{7}3$	$\bar{8}7$	$\bar{8}1$	$\bar{7}0$
$h=6$	128	121	102	76	49	27	17	20	32	50	66
$h=8$	83	82	79	74	68	60	52	42	31	21	11
$h=10$	89	85	77	63	44	23	2	$\bar{1}9$	$\bar{3}6$	$\bar{4}9$	$\bar{5}7$

Values of $y=$	$33°$	36	39	42	45	48	51	54	57	60
$h=0$	32	40	47	50	49	45	37	30	20	12
$h=2$	$\bar{3}5$	$\bar{1}6$	7	27	44	55	59	56	49	38
$h=4$	$\bar{5}7$	$\bar{4}4$	$\bar{3}5$	$\bar{3}2$	$\bar{3}3$	$\bar{4}2$	$\bar{4}9$	$\bar{5}4$	$\bar{5}3$	$\bar{4}6$
$h=6$	74	68	47	14	$\bar{2}4$	$\bar{6}0$	$\bar{8}9$	$\bar{1}02$	$\bar{9}8$	$\bar{7}8$
$h=8$	2	7	$\bar{1}3$	$\bar{2}1$	$\bar{2}6$	$\bar{2}9$	$\bar{3}1$	$\bar{3}1$	$\bar{3}0$	$\bar{2}7$
$h=10$	$\bar{5}9$	$\bar{5}6$	$\bar{4}8$	$\bar{3}4$	$\bar{1}9$	$\bar{2}$	14	28	39	47

Values of $y=$	$63°$	66	69	72	75	78	81	84	87	90
$h=0$	4	0	$\bar{3}$	$\bar{4}$	$\bar{3}$	$\bar{4}$	$\bar{6}$	$\bar{9}$	$\bar{1}3$	$\bar{1}8$
$h=2$	24	13	4	$\bar{1}$	$\bar{2}$	$\bar{1}$	4	7	11	12
$h=4$	$\bar{3}7$	$\bar{2}3$	$\bar{9}$	1	6	8	6	2	0	$\bar{2}$
$h=6$	$\bar{5}0$	$\bar{1}8$	10	27	29	19	2	$\bar{1}6$	$\bar{3}1$	$\bar{3}6$
$h=8$	$\bar{2}5$	$\bar{2}1$	$\bar{1}6$	$\bar{1}2$	$\bar{8}$	$\bar{4}$	0	3	4	5
$h=10$	49	48	43	34	23	12	1	7	$\bar{1}3$	$\bar{1}5$

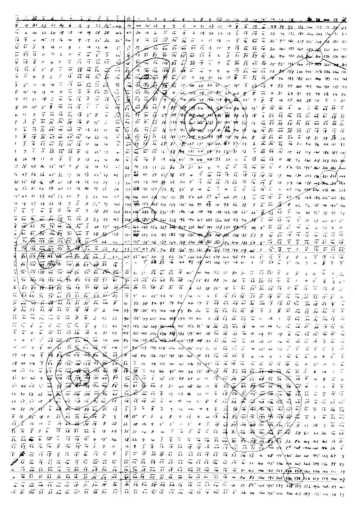

FIG. 18. Representation of the results of a two-dimensional synthesis of the neutron scattering density of α-resorcinol, $m-C_6H_4(OH)_2$ projected on the 001 plane. The grid of points has an interval of 1/120th of the cell-side (i.e. 3°) in each direction x, y, so that the figure covers a little more than a quarter of the cell edge in each direction. Contour lines are drawn to indicate the relative magnitude of some of the atomic scatterers and the atomic centres are joined, to indicate the shape of the molecule and the bond directions.

order to overcome this difficulty it is important to acknowledge that if we can postulate a model for a centro-symmetrical structure which is approximately correct and we then proceed to calculate, from equation (5.2), the expected signs for the structure factors F_{hkl}, then we shall find in due course that most of these signs are correct for the true structure. For most of the reflections, and particularly for the intense low-angle reflections, small errors in assumed atomic positions will not have a very large overall influence on the values of the signs. We can therefore concentrate our efforts on suggesting or "guessing" models of our structure which are approximately correct. The ability of a crystallographer to use every available scrap of information as a clue to the discovery of a structure which is approximately correct is quite fundamental. We shall mention here only some of the main considerations in trying to deduce a plausible structure. The reader who wishes to examine this subject in greater detail is referred to a good discussion in Bunn's *Chemical Crystallography* (1961). Some of the preliminary procedure is straightforward. In particular, the unit cell and space group are determined and the number of molecules per unit cell can then be found from a knowledge of the density. We then consider possible positions for the molecules relative to the symmetry elements. The nature of equivalent positions and their multiplicities in the particular space group must be examined: for example, an atom in a general position cannot be closer to a rotation axis or symmetry plane than its own radius. Quite general considerations and correlations between different compounds may be very informative. Thus in series of compounds, particularly homologous series of organic compounds, it is possible to make very important deductions from the unit cell dimension alone. For example, Paper 8 entitled "The Structure of Organic Crystals" was written by W. H. Bragg in 1921 and determines the basic structures of napthalene, anthracene, and some of their derivatives from very simple considerations and establishes for the first time the significance of the chemical molecule in building up structures of this kind. Once this concept was established it became clear that the occurrence of very intense reflections might indicate that molecules lay in particular planes. For example, in benzoquinone Robertson (1935) found a very intense 201 reflection, but weak reflections for 402 and 603, and was able to interpret this by placing the flat molecules nearly on the 201 planes. Other outstanding examples of this kind of feature are the intense 10, 0, 0 reflections for rubber (Bunn, 1942) and the 0, 3, 17 for chrysene (Iball, 1934).

There are, in special cases, two possible direct approaches to solving the problem set by our lack of knowledge of the sign or phase of the structure amplitude factors. The first, known as the "heavy atom" method, can be used when a compound possesses a heavy-atom derivative which is isomorphous with the parent substance and in which the heavy atom is at a known or easily determined point in the unit cell. The application of this method by Robertson to determine the structure of phthalocyanine is described in Paper 9. Phthalocyanine, $C_{32}H_{18}N_8$, has monoclinic symmetry and there are two molecules in its unit cell. It forms a nickel derivative which is closely isomorphous and has a unit cell of almost identical size, in which the nickel atoms can be shown to be located at the centres of symmetry 000 and $\frac{1}{2}\frac{1}{2}0$. For each reflection a comparison is made of the intensities observed for crystals of free phthalocyanine and the metal derivative. If it is found that the intensity of the latter is greater than that for the free phthalocyanine then it follows that the sign of F in the free phthalocyanine is the same as that from the nickel atom and we can deduce this sign from our knowledge that the nickel atom is located at the origin. In this way Robertson was able to ascribe signs directly to all the reflections in the $h0l$ zone and could then proceed to the determination of a projection of electron density in the 010 plane. This method depends, of course, on the existence of a suitable compound which includes a metal of dominant scattering power. A variation of the method applies to the case of compounds which comprise an isomorphous series in which we can determine the way in which the F value varies for metal atoms of increasing scattering power. This method can be illustrated by some data for a series of alums

$RAl(SO_4)_2 12H_2O$, where R is a monovalent ion (Cork, 1927). Table IV lists the values of F for successive hhh reflections in the cases of the alums of ammonium, potassium, rubidium and thallium. In the regions of the table which are underlined there is an increase of F value with increasing scattering power of the

TABLE IV. INTENSITIES OF hhh REFLECTIONS IN THE
ISOMORPHOUS SERIES $RAl(SO_4)_2 12H_2O$

Ion		NH_4	K	Rb	Tl
Reflection	Number of electrons	11	19	37	81
111		86	38	29	113
222		0	19	79	195
333		111	125	158	236
444		25	6	55	125
555		24	49	64	131
666		86	86	122	164
777		53	34	0	18
888		0	16	22	56
999		25	0	0	25

metal atom, leading to the conclusion that the sign of F is the same as the sign of the contribution from the metal atom. This knowledge then enables us to specify the various signs.

Apart from these procedures which can only be applied to particular groups of compounds to yield a direct method of analysis we must mention a procedure which is less powerful but, which, because of its general applicability, has been very widely used since its first conception in 1934. Known as the

Patterson synthesis, after its discoverer, it arises from consideration of a series of the same form as equation (5.4) but one in which the coefficient is not F_{hkl}, which equals C_{hkl}, but F_{hkl}^2. We shall in fact show that such a summation of F^2, although failing to give information about atomic positions, does yield information about *interatomic distances*. It can be shown that the triple summation

$$P(u,v,w) \equiv \frac{1}{V} \underset{hkl}{\Sigma\Sigma\Sigma} F_{hkl}^2 \exp 2\pi i(hu/a + kv/b + lw/c)$$

is equal to

$$V \int \int \int \rho_{xyz}\, \rho_{x+u,\ y+v,\ z+w}\, dx\, dy\, dz \qquad (5.6)$$

where V is the volume of the unit cell over which the integral is taken and ρ_{xyz} is, as before, the electron density at the point x,y,z. Equation (5.6) can therefore be used to define a Patterson function $P(u,v,w)$ as

$$P(u,v,w) = \int \int \int \rho_{xyz}\, \rho_{x+u,\ y+v,\ z+w}\, dx\, dy\, dz \qquad (5.7)$$

where the integral is taken over the unit cell. We can understand the physical significance of this function, or of the simpler two-dimensional function

$$P(u,v) = \int \int \rho_{xy}\, \rho_{x+u,\ y+v}\, dx\, dy \qquad (5.8)$$

with the aid of Fig. 19. If P is the point $x = u, y = v$ in the unit cell then the summation under the integral sign is equivalent to "scanning" the unit cell with the vector OP and, for each position such as $O'P'$, multiplying together the electron densities which exist at the two ends of the vector. It is clear that there will be a substantial contribution to the integral only if there is a concentration of electron density at *both* ends of the vector. Thus when we are calculating $P(u,v)$ there will be a contribution from $O'P'$, but not from $O''P''$ or even from $O'''P'''$. This means that if we get a large value for $P(u,v)$ then there must exist somewhere within the unit cell an interatomic separation represented by (u,v). An example of the way in which $P(u,v)$ varies when it is plotted out over the unit cell is given in Paper 10, which is

c

Patterson's original account of this procedure. It is important to emphasize that the plot of $P(u,v)$ indicates only atomic *separations* and it is not concerned with absolute atomic positions. Thus, if there are *two* pairs of atoms with a separation vector represented by the co-ordinates u, v then they will give a single, and inseparable peak. In particular, at the origin of a Patterson projection there will always be an intense peak whose amplitude is equal to the sum of the squares of the scattering factors of *all* the atoms in the

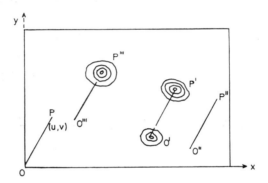

FIG. 19. Illustration of the physical significance of the Patterson function which for the point u, v is defined as the summation of $\rho_{u,v}\rho_{x+u, y+v}dx\ dy$ taken all over the unit cell. In the diagram the only large contribution to $P(u,v)$ comes from the component given by $O'P'$. Thus a large value of $P(u,v)$ for particular values u_0, v_0 indicates the presence of a pair of atoms with this *interatomic separation*.

unit cell. The usefulness of the Patterson projection in any particular structure analysis will depend very much on the particular atoms which are present in the unit cell. The pattern will be very difficult to interpret if the unit cell contains mainly atoms of roughly equal scattering power, but will give significant information if there exists one atom which is much heavier than the others. In this case the pattern of $P(u,v)$ will be very similar to a picture of electron density in the cell, seen from the position of the heavy atom: thus, in this approximate picture the origin will be the site of the heavy atom. Even in this case, however, there is one additional complication: there is no difference in value between $P(u,v)$ and $P(-u, -v)$ as defined by equation (5.8) so that

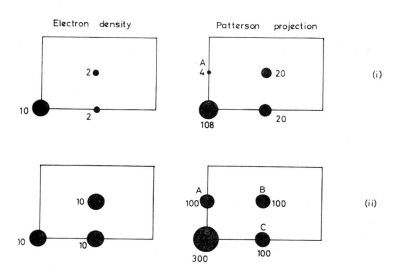

FIG. 20. A comparison of the electron density plot and the Patterson projection for a unit cell containing (i) a heavy atom and two light atoms, and (ii) three atoms of equal weight. In (ii) the peak at A, which is equal in magnitude to B and C, has no counterpart in the electron density map. On the other hand for (i), which refers to a unit cell with a heavy atom, only a very small spurious peak appears at A. In these diagrams we have shown, for simplicity, only the distributions about the bottom left-hand corner of the cell: a similar distribution belongs to *each* corner of the cell.

a centre of symmetry is always introduced artificially into the Patterson projection. The significance of the heavy atom, in producing a Patterson projection which has identifiable similarities to the *electron-density projection*, is indicated in the simple examples of Fig. 20. In (i) where there is a heavy atom the

Patterson projection is much more similar to the electron projection than is the case of example (ii) where the three atoms are of equal density. In spite of the limitations and ambiguities of the Patterson method, it remains as a very valuable contributor of leading information and is of the utmost value in the early stages of any structural analysis.

In the case of a structure which possesses a centre of symmetry the "phase problem" reduces, as we have seen, to a determination of sign. We require to know whether the sign of each structure amplitude factor F is $+$ or $-$. It may reasonably be accepted by the reader that the individual signs are not randomly independent, seeing that they are all specified once an arrangement of atoms is postulated, and it can in fact be shown mathematically that various statistical relations exist among the signs of particular reflections. Much research has been carried out on the use of statistical methods for contributing to a solution of the phase problem and these have been successful in individual cases. However they do not give a general direct solution and we shall not deal with them further here. The reader who is interested in a general account is referred to Bunn's *Chemical Crystallography* (1961) which we have mentioned earlier.

Before concluding this chapter we return to a discussion of the essentially "trial-and-error" procedure for arriving at the correct crystal structure. Using information of various kinds as we have seen, and with the likely assistance of a Patterson projection, we suggest a possible trial structure. Assuming that this structure is correct we then proceed to calculate the signs (and magnitudes) expected for each structure factor F and we use these signs to construct a Fourier projection from our *experimental* intensity data. If our assumed model is correct, then this "experimental" projection will reproduce the detail of the model which we chose. Any deviations from it, such as spurious atomic peaks, will indicate errors in the model and the nature of the deviations will enable us to suggest modifications of the model, leading to a second calculation of signs followed by a further calculation of electron density. By a continuance of this process we reach a stage at which the calculated density is in agreement with the model and this we then accept as our chosen structure. Inevitably the agreement will not be perfect and it is useful to have some index of the accuracy of our determination. This, or more strictly the *error* of the determination, is expressed by a Discrepancy Factor which we express as $\Sigma|F_0 - F_c|/\Sigma|F_c|$, where F_0, F_c are respectively the experimental value of each structure factor and that calculated from our final model. In an accurate structure analysis the value of this Discrepancy Factor, usually denoted by R, may be as low as 3 per cent. Such accuracy as this can only be obtained if we have developed a very detailed model and if we started from very accurate experimental data. A detailed model of this kind will specify not only the atomic co-ordinates of all the individual atoms in the unit cell but also the amplitudes of the vibrations about their mean positions, which they undergo because of their possession of thermal energy. This so-called "thermal motion" may be anisotropic and each atom may require six thermal parameters, as well as its three Cartesian co-ordinates, to specify its behaviour completely. Before this degree of precision has been reached in the details of a model it is usually profitable to use the "least-squares" method of refinement, rather than continuing with a succession of Fourier projections. The principle of this method, readily carried out with a computer, is to postulate small changes in parameter values and express the condition that the sum of the squares of the resulting errors, that is the difference between observed and calculated F values, should be a minimum. If an abundance of accurate intensity measurements is available then a number of successive refinements in this way will produce very accurate values of both the atomic co-ordinates and the parameters which specify the thermal motion of the atoms.

CHAPTER VI

Applications and Limitations of X-ray Diffraction

OUR main interest so far has been in the principles of X-ray diffraction and the way in which these can be applied to determine the three-dimensional structure of solids. We have seen that the techniques are the most powerful in the case of well-crystallized material, for which an individual crystal with linear dimensions perhaps between 0·1 and 1 mm can be separately identified and aligned for examination in an X-ray beam. On the other hand, although we can produce a sharp and clearly defined pattern from a powdered material it is much more difficult to utilize such a pattern for determining an unknown structure. Nevertheless, this latter pattern is very characteristic of the material, affording as it does systematic information which relates diffracted intensity and interplanar spacing and the pattern may be regarded as the "finger-print" of the particular crystalline material. In fact the production of such a pattern, using the standard Debye–Scherrer methods of photography or counter detection, provides a standard and routine method for the identification of solid materials. Information on known materials has been tabulated and systematized in the so-called A.S.T.M. index produced by the American Society for Testing Materials in such a way that the pattern of a substance, or a component in a mixture, can be identified by correlation with the classified information. Such methods of identification, constituting a primary way of non-destructive testing, might be regarded as constituting the most important application of X-ray diffraction methods apart from direct structural determinations.

Between these two very distinct applications, which may be considered as two extremes, lie the applications of diffraction to studying what we may broadly call the "texture" of materials. This term can include all types of crystal imperfection, on a coarse or microscopic scale, together with a study of crystallite size and shape and the way the crystallites may be preferentially oriented in polycrystalline fragments. Briefly, the texture enables us to describe in more detail an ordinary crystalline sample which is unlikely to be either a perfect idealized single-crystal or a completely random collection of spherical crystal grains. We must emphasize that it proves necessary to be aware of the texture of a crystal even in straightforward determinations of crystal structure. The reason for this is that the crystal texture has an important influence on the relation between the structure factors and the intensities of the corresponding diffraction spectra. Thus, for an ideal perfect crystal, in which the three-dimensional structure is absolutely perfect and uniform, it can be shown that the intensity of any spectrum is proportional to its structure factor F. However, "ordinary" crystals are not completely perfect but include dislocations and faults, giving them what is often called a "mosaic structure", and for them the intensity is in fact proportional to F^2, the *square* of the structure factor. Other factors too, such as absorption, have to be considered in any quantitative interpretation of the intensities.

It is the study of these topics relating to texture which provide an important application of X-ray diffraction in industrial and technical laboratories, for example in metallurgy where we may study the effect of heat treatment and mechanical working on the texture and the subsequent physical properties of metals and alloys. Applications of this kind have been well described in the literature and we shall not deal further with them here. The reader who is interested can consult such a compendium as the book entitled *X-ray Diffraction Analysis of Polycrystalline Materials* (Peiser, Rooksby, and Wilson, 1960), which will make clear the practical possibilities and limitations of these techniques. Here we shall pass on to consider three much more fundamental limitations in the use of X-rays for giving a complete picture of the structure of a solid and we shall then indicate how we can get a more satisfactory picture by supplementing our knowledge with experiments which use a beam of neutrons instead of X-rays.

We have noted earlier that the process whereby atoms, and solids, scatter X-rays is an interaction between the electromagnetic radiation field and the electrons of the atom, thus leading to our conception of the unit cell as a distribution of electron density. It will be clear therefore that the prominence, in this picture, of any particular atom will be directly proportional to the number of electrons that it contains, i.e. to its atomic number. Thus, an atom of lead will be eight-two times as prominent as an atom of hydrogen and the latter would be difficult to distinguish. Indeed, even in the presence only of atoms such as carbon and oxygen we shall need intensity measurements of high accuracy in order that hydrogen atoms shall be clearly depicted in the projections of electron density. The early X-ray investigations gave very little information about the co-ordinates of the hydrogen atoms and none whatever about their thermal motions. A second limitation in our X-ray picture of a structure arises because of possible discrepancies between the position of an atomic nucleus and the centre of gravity of the electron cloud which surrounds it: it is possible, for example, that the electron cloud is extended asymmetrically as a consequence of the atom being bonded to a neighbouring atom. There is evidence that this occurs to some extent for the carbon atoms in graphite, which are found to have a slight excess of electron density in the directions of their three carbon neighbours. It would therefore be of some advantage to have a technique which locates nuclei, and defines inter-nuclear distances, rather than determining electron density. The third limitation of X-ray methods, and the one which in retrospect will be seen as the most significant one, is that X-rays take no note of electron spin and are therefore not able to characterize or distinguish separately the unpaired electrons which give rise to magnetic properties in solids.

In the following chapters we shall see how the above three limitations can be removed if we use beams of thermal neutrons, instead of X-rays, for our diffraction experiments.   We shall proceed therefore to an account of how the discovery of the neutron led to an examination of the way in which atoms would scatter neutrons and, in turn, to the development of methods of using neutron beams for studying solids.

CHAPTER VII

The Diffraction of Neutrons

THE neutron was discovered by Chadwick in 1932 and was shown to be an uncharged particle with a mass approximately equal to that of a hydrogen atom. The absence of any electric charge accounted for the extraordinary ability of the neutron to penetrate solid substances. According to the wave-mechanical theory neutrons should show certain wave properties, with a wavelength λ given by the de Broglie relationship

$$\lambda = h/mv \tag{7.1}$$

where h is Planck's constant and m,v are the mass and velocity of the neutron. Thus the wavelength becomes shorter as the neutron velocity increases. If we consider the neutrons which are produced from an ordinary radium–beryllium source surrounded by paraffin wax then they will make collisions with hydrogen atoms in the paraffin and will emerge with a root-mean-square velocity v such that

$$\tfrac{1}{2}mv^2 = 3/2 \, kT \tag{7.2}$$

where k is Boltzmann's constant and T is the absolute temperature. A combination of equations (7.1) and (7.2) shows that the neutrons will have wavelengths lying predominantly between 1–2 Å, so that their wavelengths are roughly equal to the interplanar spacings of ordinary crystalline substances. We should expect, in principle, therefore that it would be possible to diffract neutrons in the same kind of way that we have been using X-rays for studying crystalline solids.

At the present time, when relatively intense beams of neutrons are available from nuclear reactors, it is easy to demonstrate directly that neutrons can be diffracted. We can produce a well-collimated beam of what we call "monochromatic" neutrons, i.e. neutrons which all have the same velocity, and allow it to fall upon a crystal. We can then show that diffraction takes place only for discrete angles of incidence, given by the Bragg equation $\lambda = 2d \sin \theta$, and we can collect the diffracted neutrons in a suitable detecting counter. The experiment is very similar to the original experiments which the Braggs performed with the X-ray ionization spectrometer. However, the neutron sources which were available to experimenters in the 1930's were very weak, far too weak to do a refined experiment in which an initial selection of monochromatic neutrons was made or in which any good angular collimation of the beam was achieved. In 1936 Elsasser considered the way in which a beam of thermal neutrons which contained a wide range of velocities would be diffracted by a powder. As described in Paper 11, he showed that there should be a *minimum* angle, determined by the maximum nterplanar spacing of the solid and the peak wavelength of the thermal spectrum, below which no scattering would take place. Very shortly afterwards the truth of this conclusion was demonstrated experimentally by Halban and Preiswerk who reported their results as follows:

Experimental Proof of Neutron Diffraction
by H. von Halban jun. and P. Preiswerk

reported at a meeting of the Academy of Science of Paris on 6th July 1936 and published in *Comptes Rendus Acad.-Sci. Paris* **203**, 73 (1936).

In considering, in terms of Louis de Broglie's theory, the wave associated with a corpuscular neutron, Elsasser (1936) has suggested that a pattern analogous to that of Debye and Scherrer* could be obtained even with un-monochromatized neutrons.

Note by G. E. B. It is rather misleading to regard this as equivalent to a "Debye–Scherrer pattern" since this term is now universally restricted to a pattern with monochromatic radiation and in which the scattering at any chosen angle can be identified unambiguously with a particular interplanar spacing. The pattern which is being discussed in the present paper can be considered as the integrated sum of a large series of Debye–Scherrer patterns.

(a) The distribution of neutron velocities from a source of slow neutrons (radium and beryllium, surrounded by paraffin) shows a maximum for neutrons of thermal energy, with an approximately Maxwellian distribution.

(b) The absorption of thermal neutrons in the nuclei of certain detecting elements takes place with a cross-section which is inversely proportional to the neutron velocity.

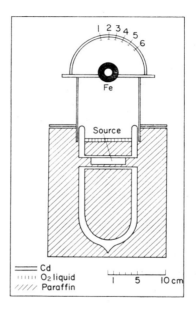

FIG. I.

Elsasser has calculated the angular distribution of the neutrons from a source which satisfies condition (a) and which are scattered by a poly-crystalline powder, assuming that they are diffracted because of their wave nature. The calculations show in particular that, for a powder of iron, the coherent neutron scattering, which can be observed using a detector which satisfies condition (b), is negligible within a cone of semi-vertical angle 26° about the direction of incidence. This angle depends on the average velocity of the neutrons and varies from a value of 18° when the moderating paraffin has a temperature of 300°K to 33° for a temperature of 90°K. The experiment consists of observing the angular distribution of the neutrons for these two temperatures.

Figure I shows the experimental arrangement. A 1 curie radium-beryllium source is placed in a Dewar flask which is filled and surrounded with paraffin. Two cadmium slits produce a collimated beam of neutrons with an angular divergence of 27°. A cylinder of iron 3 cm in height which acts as the scatterer is placed at the second slit and detectors with a surface area of 0·5–2 cm² are placed at intervals of 13° around the iron cylinder in a cadmium chamber. We have chosen dysprosium as the detecting material

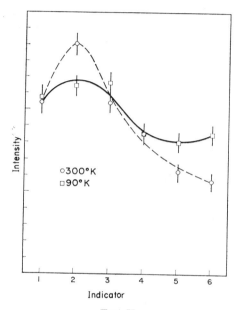

FIG. II.

because it is very sensitive to thermal neutrons. On the other hand cadmium absorbs all the neutrons which are captured by dysprosium so that there is a suitable means of completely protecting the detectors. However, the observed effect will be reduced since it can be shown by variation of temperature that the cross-section of dysprosium for thermal neutrons varies less quickly than according to $1/v$.

The observed results are shown, with their probable statistical errors, in the two curves of Figure II. When no scatterer is in position, it is found that the distribution is independent of the temperature of the paraffin. The curves show that the angular distribution from the scatterer depends on the velocity of the neutrons. When the velocity of the neutrons is reduced, the scattered intensity at small angular deviations gets less, but for larger deviations it increases. The effect is a small one but, being outside statistical error, it gives clear evidence of the diffraction effect for the wave associated with the neutrons. Quantitatively we cannot compare these results with Elsasser's calculations: in our experiment the sensitivity of the detectors varies less quickly than $1/v$ and the incident beam necessarily needs a large angular divergence because of the weak neutron intensity of the source.

About a month later the result of a more refined experiment was reported in the *Physical Review* by Mitchell and Powers. A similar source of neutrons was used but diffraction took place from a set of large single crystals, of magnesium oxide, and the diffracted beam was measured directly—in contrast with the experiment of Halban and Preiswerk which really measures the sum of the direct and diffracted beams. Mitchell and Powers's report of their experiment appears in Paper 12.

Although these two experiments demonstrated very satisfactorily that neutrons could be diffracted by solids, no applications of these methods to the actual study of solids were possible so long as only very weak sources of neutrons were available. The situation was transformed around 1945 when the nuclear reactor had been developed and, for the first time, relatively abundant quantities of neutrons became available. Paper 13 by Zinn describes the first "neutron spectrometer" by means of which neutrons with a narrow band of wavelength are selected from the Maxwell spectrum which emerges from the reactor. The selection was achieved by a large single crystal, functioning as a mono-chromator, whose angular position determined the wavelength of the diffracted monochromatic beam. The beams produced in this manner were used for a quantitative study of the way in which individual atomic nuclei scattered neutrons and how this depended on the neutron wavelength and the existence of "resonances" in the atoms. Zinn's paper describes only the principal features of this work. It was accompanied by a paper by Sturm (1947), not reproduced here, but which gives many more experimental details and scattering data for many elements, and by a most important paper by Fermi and Marshall, reproduced as Paper 14. The latter paper may be regarded as the one which first revealed to physicists and chemists the prospects of using neutron beams for studying solids. In particular the paper makes clear how the scattering amplitude of a nuclear species can be determined and it discusses how this amplitude may vary for different isotopes and how it depends on nuclear spin. At the same time it establishes, by experimental measurements using mirror reflection, that for most nuclei there is a phase change of 180° when neutrons are scattered, but for a few exceptional elements—Li, Mn, H—the phase change is zero. In the following years great progress was made at the Oak Ridge National Laboratory in the U.S.A., culminating in the appearance of a paper by Shull and Wollan (1951) which, amongst other things, listed the coherent scattering amplitudes for about sixty elements and isotopes. This work may be regarded as having established the potential of neutron diffraction as a technique and tool for studying solids. Henceforth it became customary to speak of the "applications" of neutron diffraction.

For clarity we shall pause here in following our path through the classical literature of neutron diffraction to emphasize the fundamental principles of the neutron-scattering process which have emerged, particularly in so far as they lead to the removal of some of the basic limitations of X-ray diffraction which we formulated in our previous chapter. It will be clear that, to the extent that we have discussed it so far, the process whereby atoms scatter neutrons is a *nuclear* one, in contrast with the electronic scattering of X-rays. It is found that the quantitative value of the scattering, i.e. the amplitude of the scattered neutron wave or "scattering length", depends on two factors. First of all, the scattering increases with the nuclear size, and there is a contribution to the scattering length proportional to $A^{\frac{1}{3}}$ where A is the mass number of the nucleus. This contribution is called "potential scattering". Secondly, there may be "resonance scattering" depending on the nuclear energy levels or, more specifically, on the energy levels in the "compound" nucleus composed of the initial nucleus and the neutron. The energies of such levels will depend on the detailed nuclear structure so that this resonance effect will vary widely from element to element, and indeed from isotope to isotope. The result is that the overall scattering varies considerably, in a haphazard way, as we go through the Periodic Table of elements. Some of the data are summarized in Fig. 21, in which the linear variation of scattering amplitude for X-rays is also indicated. It will be noticed that although the random variations from element to element are substantial for neutrons,

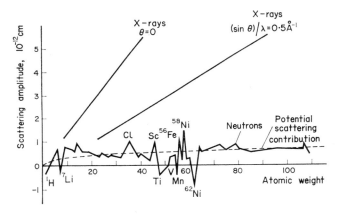

FIG. 21. The variation of the scattering amplitude of atoms for neutrons as a function of their atomic weight. The irregular variation is contrasted with the linear variation for X-rays, which depends also on the angle of scattering (from *Research* (1954) **7**, 297).

nevertheless the magnitude of the overall variation is comparatively small. The average value, taken over all the elements and isotopes which have been measured so far, is about 0.6×10^{-12} cm and there are very few values which are less than a half of this average or greater than twice it.

It follows therefore that if we employ with neutrons the same diffraction procedures that we have discussed for X-rays then we shall be able to obtain three-dimensional pictures, or two-dimensional projections, of *nuclear* scattering density, thus revealing nuclear positions instead of electron clouds. In these pictures hydrogen will show up substantially and will have no great inferiority compared with heavy elements. The scattering length of hydrogen, -0.38×10^{-12} cm, is a little below the average value for all nuclei but in cases of difficulty we can produce very favourable conditions for detection and location by using deuterated material, instead of samples containing ordinary hydrogen; the scattering length of deuterium is 0.65×10^{-12} cm so that it is a "better-than-average" element.

A gratuitous advantage which neutrons often offer to us is the ability to distinguish between neighbouring elements, such as iron and cobalt or cobalt and nickel. This is a problem of some importance in many studies of alloys, and one which cannot be solved satisfactorily with X-rays because of the closely similar scattering amplitudes of these elements for X-rays. It is a fortunate circumstance that for these, and some other important pairs of neighbours, the neutron-scattering amplitudes are considerably different.

It is worth while to comment briefly at this stage on the technique of *electron* diffraction, since it offers certain of the advantages of neutron diffraction so far as the detection of light elements is concerned. Whereas X-ray diffraction measures electron density, electron diffraction measures *electric potential*. The electric potential has a peak at the centre of each atom and it increases in magnitude with the atomic number of the atom, but not so quickly as does the electron density. Vainshtein (1964), in a recent review of the study of light atoms by electron diffraction, has calculated the relative detectabilities of lead and oxygen atoms for X-rays and electrons. The ratio is 20 for X-rays and 6 for electrons; for neutrons the ratio would be about $1\frac{1}{2}$. Thus electrons have an advantage for light-atom detection, compared with X-rays, but they are inferior to neutrons. They are, of course, very readily available, compared with neutrons, and can be used for examining samples in the form of thin films. This is possible because of the large scattering amplitude of atoms for electrons, which is of the order of 10^{-8} cm, compared with 10^{-12} for neutrons and 10^{-12}–10^{-11} for X-rays. This means that electron diffraction patterns can be obtained from films of 100–1000 Å in thickness and that the technique is particularly valuable for studying surface structures. In other respects the restriction of the measurements to thin layers of material is a very awkward one.

In our discussion of X-ray diffraction techniques in Chapter 4 we commented that the study of single crystals was much more informative, and the results were much easier to interpret, than the use of powdered material and the Debye–Scherrer technique. We must therefore give some explanation of why all the quantitative work with neutrons prior to about 1951 was done with powdered materials. We find, for instance, that all the experimental measurements which yielded Shull and Wollan's (1951) list of the scattering amplitudes for sixty different nuclei were made with powders. There are two reasons for this. First, as we have already mentioned, the powder patterns of cubic substances are *not* difficult to interpret and for all these early basic measurements of the elements suitable compounds with simple structures of cubic symmetry could be found. Secondly, the early experimenters found unexpected discrepancies in trying to interpret their measurements of single crystals. We have stated in Chapter VI, when we mentioned the "texture" of crystals, that most crystals are faulted and "mosaic", with the consequence that reflection intensities are proportional to F^2, rather than simply to F as they would be for a perfect crystal. In this respect neutron diffraction did not appear to follow the expected rules and Fermi and Marshall, when seeking to interpret their results for single crystals, remark that "we have found that one obtains a much better fit by using the form factor", i.e. by assuming proportionality to F rather than F^2. There is a straightforward explanation of this behaviour, although a detailed quantitative treatment of the problem is extremely complicated. The difference in behaviour for neutrons arises because of their very low absorption coefficients, which means that when a neutron beam travels through a solid then the process of scattering which builds up the diffracted beams is responsible for practically all of the attenuation which takes place. With X-rays, on the other hand, fluorescent absorption accounts for the removal of most of the intensity of the incident beam and the scattering process plays a much less significant part. The practical result is mainly that for intense neutron reflections the proportionality between intensity and F^2 is only achieved for crystals up to a millimetre or so in thickness. For thicker crystals the intensity falls away from this proportionality. A fairly detailed treatment of this behaviour was given by Bacon and Lowde (1948) and an experimental justification of the arguments by Bacon (1951). The latter paper is reproduced as Paper 15. The conclusions of this paper, reinforced by a slightly later paper of Peterson and Levy (1951) which presented a comparison of intensity data for powders and single crystals, may be considered to have reoriented neutron diffraction along the path of structural analysis using single-crystals. "Neutron crystallography" was now an established subject and in the following year Peterson and Levy (1952) presented the first single-crystal analysis, using Fourier synthesis methods for interpreting their measurements for KHF_2: their account of this work appears as Paper 16. This paper caused great interest among physicists and chemists who were occupied with the study of hydrogen bonds and led quickly to a succession of papers which, with results of steadily increasing accuracy, showed the power of neutron diffraction methods in examining hydrogen bonds and molecular structure.   Reviews of this work may be consulted in the literature (Bacon (1959), Hamilton (1962), Rundle (1964)). The present status and possibilities of the method can be indicated by noting that in a recent study of sucrose, $C_{12}H_{22}O_{11}$, by Brown and Levy (1964) 5800 neutron reflections were measured and led to the determination of the co-ordinates and parameters of the forty-five atoms in the molecule. The data were collected on a diffractometer controlled and programmed by a punched tape which automatically aligned the crystal and detector in three dimensions and then scanned in succession through the individual reflections. This is the largest molecule and the most complicated structure which has been examined so far. At the other end of the scale of molecular size we have the study of the uranium oxide system by Willis (1964) in which very accurate measurements of intensity led to precise conclusions about the nature of defects and the anisotropy of thermal motion in this series of oxides which is based on the simple fluorite type of structure.

CHAPTER VIII

Magnetic Scattering of Neutrons

WHEN we began our discussion of the way in which atoms scatter neutrons we stated that, in general, the scattering process consisted of an interaction between the neutron and the *nucleus* of the atom. The qualification of this statement can be seen when we examine the scattering by magnetic materials. The neutron has a spin and an associated magnetic moment and it was pointed out by Bloch in 1936 that if the value of this magnetic moment was about the same order as the known measured moment of the proton, then there would be observable "magnetic scattering" of neutrons by those atoms which possess a resultant magnetic moment. Bloch's original letter to the editor of *Physical Review* appears as Paper 17. Experimental measurement by Alvarez and Bloch (1940) showed that the neutron moment was about 0·7 of the proton value, so that magnetic scattering comparable in value to the ordinary nuclear scattering might be expected from atoms which possess a resultant magnetic moment. A detailed discussion of the magnitude of this scattering was given in 1939 by Halpern and Johnson. This paper was of the greatest importance in guiding the future development of the subject, but is rather too detailed and lengthy for our present purpose to be reproduced in full. We shall restrict ourselves therefore to printing, as Paper 18, the abstract and the final section which suggested various experiments which could test the theoretical conclusions.

The essential feature of the magnetic scattering is that for an atom with a spin quantum number of S there will be a differential scattering cross-section $d\sigma/d\omega$, where ω is the measure of solid angle, such that

$$d\sigma = \frac{2}{3} S(S+1)\left(\frac{e^2\gamma}{mc^2}\right)^2 f^2 \, d\omega \qquad (8.1)$$

where γ is the magnetic moment of the neutron, in units of the nuclear magneton $eh/4\pi mc$, and f is a form factor. This form factor takes into account the fact that the linear dimensions of the space occupied by the magnetic electrons in the atom will be comparable with the neutron wavelength, so that the amplitude of the resultant scattered contribution will become less as the angle of scattering increases. e, m in the formula are the charge and mass of the electron and c is the velocity of light. If, an as example, we substitute a value of unity for S in the formula we shall find that the magnetic scattering in the forward direction is about equal to the nuclear scattering from an average atom.

The original suggestions of Bloch were published in 1936 and over the next few years many attempts were made to demonstrate the existence of this paramagnetic scattering. These experiments were not successful, largely because of the weakness of the radium-beryllium sources, but significant confirmation of the theoretical conclusions was obtained in some later work by Ruderman (1949) who used monochromatic neutrons from a cyclotron. The measurements showed an excess of scattering which was in good agreement with calculations based on Halpern and Johnson's formula.

With the availability of intense beams of neutrons from nuclear reactors it became possible to make well-defined experimental measurements and to use neutron beams for determining the "magnetic structures" of solids, as a development of our knowledge of what we might regard as merely the "chemical structure", secured from X-ray data. Our problem is now not merely to determine the co-ordinates in the unit cell for each atom, but to find out also (in the case of the magnetic atoms) the direction and magnitude of the magnetic moments which they carry. We can illustrate this by the simple diagrammatic arrangements shown in Fig. 22, where for simplicity we have restricted ourselves to a linear array of atoms. In (i), for a paramagnetic material, the magnetic moments lie in random directions; in (ii), which is ferromagnetic, they are all aligned upwards; in (iii), which is the

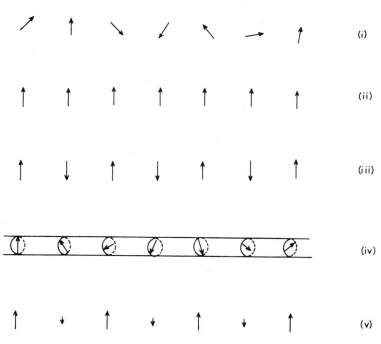

Fig. 22. Diagrammatic representations of different types of magnetic structure, illustrated for linear arrays of atoms as follows: (i) paramagnetic, (ii) ferromagnetic, (iii) simple antiferromagnetic, (iv) spiral antiferromagnetic, (v) ferrimagnetic.

simplest type of antiferromagnetic, they are directed alternately up and down; whereas in (iv), which is a more complicated antiferromagnetic structure, the direction of the magnetic moments spirals around steadily as we advance from atom to atom from left to right.   Finally, in (v) we depict a ferrimagnetic arrangement in which *unequal* moments are directed up and down. When these ideas are extended to three dimensions a wide variety of antiferro- and ferrimagnetic structures becomes possible and the details of the magnetic scattering pattern for neutrons will enable us to determine the details of these arrangements.   The first experimental determination of such a structure is described in Paper 19 by Shull and Smart who were able to deduce the arrangement of the magnetic spins in manganous fluoride MnO below its Néel temperature of 122°K.   The arrangement finally chosen after further work on this substance by later investigators was that indicated in Fig. 23.   It is clear from this figure that, from a magnetic point of view, each side of the unit cell is twice as large as for the ordinary "chemical" cell.   This accounts for the appearance of the characteristic additional lines, due solely to magnetic scattering, which appear in the upper diffraction pattern in Paper 19.   If indexed in terms of the chemical cell then these lines would be identified as, for example, $\frac{1}{2}\frac{1}{2}\frac{1}{2}$: they do not, of course, appear in X-ray diffraction patterns for these materials and this provides another means of distinguishing them and then demonstrating their significance.

The original paper by Shull and Smart was the signal which generated enormous interest in the application of neutron diffraction to studying magnetic architecture, so much so that it is now this application of the technique which is mostly widely studied.   This outcome is not surprising, since neutrons are so unique in their ability to give this direct evidence of co-operative magnetic arrangements.   Two very important papers appeared shortly after the first publication.   Shull, Strauser, and Wollan (1951) developed the early work on MnO by studying the series of oxides MnO, FeO and NiO and α-Fe_2O_3 hematite, which is also antiferromagnetic.   In the succeeding paper Shull, Wollan, and Koehler (1951) examined both ferromagnetic materials, such as metallic iron and cobalt, and magnetite which is the simplest example of a ferrimagnetic material.   They were also able to demonstrate directly the effect of a magnetic field on the intensity of neutron scattering, as Halpern and Johnson had predicted theoretically, and to produce a beam of polarized neutrons by

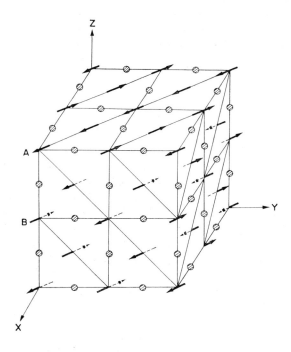

FIG. 23. The magnetic structure of MnO indicating the doubled unit cell dimensions when the magnetic moment directions are taken into account. The arrows drawn at the positions of the manganese ions Mn^{2+}, such as at A and B, indicate the directions of their magnetic moments: the small circles show the intervening oxygen atoms. As shown, the structure consists of ferromagnetic sheets of atoms parallel to the (111) plane and the moment directions lie in this plane.

reflection of their initial neutron beam by a suitable magnetized crystal. The paper describing this work is reproduced as Paper

20. The use of a beam of polarized neutrons as described in this paper has led to a great increase in the accuracy of measuring magnetic moments and, in turn, of studying the distribution of magnetic scattering power in solids. The full advantages of the method can only be secured for ferro- and ferrimagnetic materials, in which it is possible to control the moment direction in the crystal by applying a magnetic field. In ferromagnetic substances it becomes possible, as Shull and Yamada (1962) have shown, to produce a planar projection of the magnetic moment density in a solid, in a similar way to that in which X-ray diffraction leads to a projection of the total electron density in a solid.

By 1952 it was clear that the use of neutron beams had established directly the truth of the models of antiferromagnetic and ferrimagnetic solids which had been first postulated by Néel. Since then, it has been shown experimentally that these arrangements are no more than the simplest type of co-operative arrangement and that a variety of much more complicated structures exists. For example, Herpin, Meriel, and Villain (1959) showed that spiral antiferromagnetic structures could occur: this was the type of arrangement which we showed diagrammatically in Fig. 22 (iv). Thus, in Au_2Mn the direction of the magnetic moment on the manganese atoms turns in spiral fashion from plane to plane, advancing by $51°$ as we progress along the c axis from one atomic plane to the next. In some of the rare-earth metals similar structures have been found, together with other phases in which there are variations in the effective magnitude of the magnetic moment from plane to plane. Yet another type of antiferromagnetic arrangement is the "umbrella structure" suggested by Corliss, Elliot, Hastings, and Sass (1961) for CrSe in which the magnetic moment directions are not collinear but lie around the curved surface of a cone. Among the rare-earth metals there have been found *ferrimagnetic* structures with similar characteristics. In principle the details of all these structures can be deduced from the neutron diffraction patterns, although in practice difficulties arise because of the same "phase problem" which we discussed in connection with X-ray analysis.

Moreover, in the case of these magnetic substances there are often substantial difficulties in producing single crystals: indeed, in order to solve the structure in a direct way with neutrons it is necessary to have not just a single crystal but a sample which is a single magnetic domain.

CHAPTER IX

Inelastic Neutron Scattering

IN THE foregoing chapters, and in the original papers to which we have referred, we have traced the growth of the techniques of X-ray and neutron diffraction for studying the three-dimensional structure of solids. With the aid of neutrons we have extended our conception of structure to include not only a knowledge of atomic positions but also details of the magnitude, position and orientation of magnetic moments. We have emphasized what we may call the "crystallographic" approach to these problems because we have been particularly interested in seeing how the use of neutrons has complemented the classical work with X-rays. This indeed has been one of our main aims in compiling this collection of papers from the literature of the subject. However, it may be worth while in this final chapter to indicate, for completeness, a completely different approach to the study of solids and liquids which has developed with the use of neutron beams.

In our earlier discussions we have considered the way in which a regular assembly of atoms scatters a beam of X-rays or neutrons which falls upon it. We have considered the geometrical conditions which determine in what particular directions in space there will be a reinforcement, by superposition, of these scattered contributions. In these particular directions we have spoken of "diffracted beams" and "reflection planes". Only a portion of the radiation scattered by the individual atoms takes part in the co-operative reinforcement which produces these beams. We have indeed only been considering that part of the radiation which is scattered by an atom without any change of wavelength.

It is in fact possible for energy interchange between the crystal and the radiation to take place for both X-rays and neutrons. In these cases the process of scattering is "inelastic" and the energy which is exchanged between the radiation and the atom, which is not free to move but at least partially bound at some precise location, is in turn passed on to the crystal as a whole. Alternatively, energy is removed from the crystal as a whole and used to increase the energy of the neutron or X-ray photon. By studying the details of the way in which quanta of energy are exchanged with the vibrations in the crystal it is possible to evaluate the spectrum of these vibrations, and such an evaluation should be possible in principle from experiments made with either X-rays or neutrons.   If, however, we consider the process quantitatively we shall find that neutrons have an enormous advantage. We can see this if we calculate for X-rays and neutrons, in turn, the energy which corresponds to 1 Å, which is about the wavelength which we use to study the structures of solids. These energies are equal to about $10^{18}h$ and $10^{13}h$ respectively, where h is Planck's constant. Calculation of the magnitude of a quantum of crystal-vibration energy, i.e. a phonon, gives a value roughly equal to $10^{13}h$, the same value as the energy quantum for a neutron.   This means that the gain or loss of a quantum of crystal energy by a neutron will produce a very large percentage change in the neutron energy and, therefore, a large change of wavelength.   By studying the spectrum of these "inelastically scattered" neutrons we can deduce the details of the "dispersion law" for the solid, i.e. the relation between wavelength and frequency for the vibrations which can be transmitted in any chosen direction.   The study of this aspect of neutron scattering has developed very rapidly over the past ten years as nuclear reactors able to provide much more intense beams of neutrons have become available.   Good accounts of the essential principles of this work, suitable for preliminary reading, have been given by Buras and O'Connor (1959) and by Brugger (1962). Details of many of the experimental investigations which have been carried out during the succeeding years can be found in the reports of a number of international conferences which have been held on this topic, such as *Inelastic Scattering of Neutrons in Solids and Liquids* (1961) and (1963). The prospects of further development by using even more intense sources of neutrons in the future are very promising.

References

ALVAREZ, L. W. and BLOCH, F. (1940) *Phys. Rev.* **57**, 111.

BACON, G. E. (1951) *Proc. Roy. Soc.* A **209**, 397   PAPER 15

BACON, G. E. (1959) *Hydrogen Bonding*, ed. HADZI, D. Pergamon Press, Oxford, p. 23.

BACON, G. E. and LOWDE, R. D. (1948) *Acta Cryst.* **1**, 303.

BERNAL, J. D. (1927) *Proc. Roy. Soc.* A **113**, 117.   PAPER 5

BLOCH, F. (1936) *Phys. Rev.* **50**, 259.   PAPER 17

BRAGG, W. H. (1921) *Proc. Phys. Soc.* **34**, 33.   PAPER 8

BRAGG, W. L. (1913) *Proc. Camb. Phil. Soc.* **17**, 43.   PAPER 2

BRAGG, W. L. (1914a) *Phil. Mag.* **28**, 355.   PAPER 3

BRAGG, W. L. (1914b) *Proc. Roy. Soc.* A **88**, 428.

BRAGG, W. L. (1929) *Z. Krist.* **70**, 475.   PAPER 6

BROWN, G. M. and LEVY, H. A. (1964) *J. Phys.* **25**, 469.

BRUGGER, R. M. (1962) *Int. Science and Technology*, p. 52.

BUNN, C. W. (1942) *Proc. Roy. Soc.* A **180**, 40.

BUNN, C. W. (1961) *Chemical Crystallography*, Clarendon Press, Oxford, 2nd edition.

BURAS, B. and O'CONNOR, D. (1959) *Nukleonika* **4**, 119.

COMPTON, A. H. and ALLISON, S. K. (1935) *X-rays in Theory and Experiment*, Van Nostrand, New York, p. 56.

CORK, J. M. (1927) *Phil. Mag.* **4**, 688.

CORLISS, L. M., ELLIOTT, N., HASTINGS, J. M. and SASS, R. L. (1961) *Phys. Rev.* **122**, 1402.

DEBYE, P. and SCHERRER P. (1916) *Phys. Z.* **17**, 277.

ELSASSER, W. M. (1936) *C. R. Acad. Sci. Paris* **202**, 1029.   PAPER 11

EWALD, P. P. (1962) ed., *Fifty Years of X-Ray Diffraction*, Oosthoek, Utrecht.

FERMI, E. and MARSHALL, L. (1947) *Phys. Rev.* **71**, 666.   PAPER 14

FRIEDRICH, W., KNIPPING, P. and LAUE, M. (1912) *Proc. Bavarian Acad. of Science*, p. 303.   PAPER 1

HALBAN, H. and PREISWERK, P. (1936) *C.R. Acad. Sci. Paris* **203**, 73.
INCLUDED IN CHAPTER VII

HALPERN, O. and JOHNSON, M. H. (1939) *Phys. Rev.* **55**, 898.   PAPER 18

HAMILTON, W. C. (1962) *Annual Reviews of Phys. Chemistry* **13**, 19.

HERPIN, A., MERIEL, P. and VILLAIN, J. (1959) *C.R. Acad. Sci. Paris* **249**, 1334.

HULL, A. W. (1917) *Phys. Rev.* **10**, 661.   PAPER 4

IBALL, J. (1934) *Proc. Roy. Soc.* A **146**, 140.

Inelastic Scattering of Neutrons in Solids and Liquids (1961) published by I.A.E.A., Vienna.

Inelastic Scattering of Neutrons in Solids and Liquids (1963) published by I.A.E.A., Vienna.

JENKINS, F. A. and WHITE, H. E. (1957) *Fundamentals of Optics*, McGraw Hill, New York.

LIPSON, H. and BEEVERS, C. A. (1936) *Proc. Phys. Soc.* **48**, 772.   PAPER 7

MITCHELL, D. P. and POWERS, P. N. (1936) *Phys. Rev.* **50**, 486.   PAPER 12

PATTERSON, A. L. (1934) *Phys. Rev.* **46**, 372.   PAPER 10

PEISER, H. S., ROOKSBY, H. P. and WILSON, A. J. C. (1960) ed., *X-Ray Diffraction by Polycrystalline Materials*, Chapman and Hall, London.

PETERSON, S. W. and LEVY, H. A. (1951) *J. Chem. Phys.* **19**, 1416.

PETERSON, S. W. and LEVY, H. A. (1952) *J. Chem. Phys.* **20**, 704.   PAPER 16

PHILLIPS, F. C. (1963) *An Introduction to Crystallography*, 3rd ed., Longmans, London.

ROBERTSON, J. M. (1935) *Proc. Roy. Soc.* A **150**, 106.

ROBERTSON, J. M. (1936) *J. Chem. Soc.*, p. 1195.   PAPER 9

RUDERMAN, I. W. (1949) *Phys. Rev.* **76**, 1572.

RUNDLE, R. E. (1964) *J. Phys.* **25**, 487.

SHULL, C. G. and SMART, J. S. (1949) *Phys. Rev.* **76**, 1256.   PAPER 19

SHULL, C. G., STRAUSER, W. A. and WOLLAN, E. O. (1951) *Phys. Rev.* **83**, 333.

SHULL, C. G. and WOLLAN, E. O. (1951) *Phys. Rev.* **81**, 527.

SHULL, C. G., WOLLAN, E. O. and KOEHLER, W. C. (1951) *Phys. Rev.* **84**, 912.   PAPER 20

SHULL, C. G. and YAMADA, Y. (1962) *J. Phys. Soc. Japan* **17**, Suppl. BIII,1.

STONE, J. M. (1963) *Radiation and Optics*, McGraw Hill, New York.

STURM, W. J. (1947) *Phys. Rev.* **71**, 757.

VAINSHTEIN, B. K. (1964) *Advances in Structure Research by Diffraction Methods*, ed. Brill, Vieweg, Braunschweig, Vol. I, p. 24.

WILLIS, B. T. M. (1964) *J. Phys.* **25**, 431.

ZINN, W. H. (1947) *Phys. Rev.* **71**, 752.   PAPER 13

PART II

Extracts

1. X-ray Interference Phenomena*

W. FRIEDRICH, P. KNIPPING and M. LAUE

Theoretical Part—M. Laue

Introduction

The investigations of Barkla[1] in recent years have shown that X-rays undergo scattering when passing through material substances, corresponding closely to the scattering of light in turbid media, but that in addition the radiation excites the atoms of the substance, causing them to emit a spectroscopically homogeneous characteristic radiation (fluorescence radiation), this radiation being characteristic of the emitting substance.

As early as 1850 Bravais introduced into crystallography the theory that the atoms of a crystal are arranged in a spatial lattice. If X-radiation truly consists of electromagnetic waves, it would be expected that when the atoms are excited, to produce free or forced oscillations, then the spatial lattice structure would give rise to interference phenomena; it would also be expected that these phenomena would be of the same nature as the lattice spectra already known in optics. The constants of such lattices may be estimated from the molecular weight of the crystallized compound, its density, the number of molecules per gram-molecule and crystallographic data. It is always found to be of

*Presented by A. Sommerfeld to the Session held on 8 June 1912. From the *Proceedings of the Bavarian Academy of Sciences*, 1912, pp. 303 ff. Reproduced by permission of Professor Theodore von Laue, Washington University, St. Louis.

[1] G. C. Barkla, *Phil. Mag.* **22**, 396 (1911).

the order of magnitude of 10^{-8} cm, while the wavelength of X-radiation, according to the diffraction experiments of Walter and Pohl,[2] and according to the work of Sommerfeld and Koch[3] are of the order of 10^{-9} cm. Investigations of X-rays are subject to the considerable complication that a spatial lattice has a threefold periodicity, while in optical lattices only one direction, or (in the case of cross-lattices) at most two, exhibit periodicity.

Friedrich and Knipping have at my suggestion submitted this possibility to experimental test. The experiments and the results obtained from them will be dealt with in the second part of this publication.

The theory and its qualitative comparison with experiment

We shall now attempt to give mathematical form to the suggestion made above. The location of the centre of an atom is determined by means of the rectangular coordinates x, y, z, whose axes intersect in an origin located at the centre of some arbitrary atom in the irradiated portion of the crystal. The spatial lattice considered is the most general one, that is, the triclinic type. The edges of the elementary parallelepiped of this crystal may thus have any lengths and be inclined at any angles to one another. By particular choice of these lengths and angles it is always possible to move over to spatial lattices of other types. We designate these edges with respect to length and direction by means of the vectors a_1, a_2, a_3, and so locate the centre of an atom in a position given by:

$$\begin{aligned}
x &= m(a_1)_x + n(a_2)_x + p(a_3)_x \\
y &= m(a_1)_y + n(a_2)_y + p(a_3)_y \\
z &= m(a_1)_z + n(a_2)_z + p(a_3)_z
\end{aligned} \qquad (1)$$

[2] R. Walter and R. Pohl, *Ann. der Phys.* **25,** 715 (1908); **29,** 331 (1908).

[3] A. Sommerfeld, *Ann. der Phys.* **38,** 473 (1912); P. P. Koch, *Ann. der Phys.* **38,** 507 (1912).

in which m, n and p may be positive or negative whole numbers, except zero, designating the atom in question.

For the oscillation of an individual atom we shall first make the assumption that it has a pure sinusoidal form. Obviously this is not strictly correct, any more than it would be in optics. All the same, just as in optics, it is possible to represent spectroscopically non-homogeneous radiation by means of a Fourier analysis as the summation of sinusoidal oscillations. The waves emitted from an atom may then be represented at large distances from the atom by means of the expression.

$$\frac{\psi e^{-ikr}}{r} \tag{2}$$

where r represents the magnitude of the radius vector from atom to point of observation, ψ is a function of its direction, and $k = (2\pi)/\lambda$, where λ is the wavelength of the X-radiation subsequently to undergo interference. If the atom were small in relation to the wavelength, as is the case in optics, then ψ would be constant.[4] We have here, however, to reckon with the possibility, which experimental evidence renders likely, that differences depending on direction will arise because of the comparable dimensions of wavelength and atom. If we further assume that the exciting oscillation moves in waves having the velocity of light, as assumed above, then we shall see that it is also necessary to add the factor $\exp(-ik(x\alpha_0 + y\beta_0 + z\gamma_0))$, where α_0, β_0 and γ_0 are the

---

[4] Commencing with any number of spherical waves from radiation sources lying in a finite region, the excitation at a point for which the radius vector from the mid-point of the region has the value r which is large in comparison with the dimensions of this region, and the direction cosines α, β, γ is:

$$\sum^{n} \frac{e^{-ik(r_n - \theta_n)}}{r_n} = \frac{e^{-ikr}}{r} \cdot \sum^{n} e^{ik(x_n \cdot \alpha + y_n \cdot \beta + z_n \cdot \gamma) - i\theta_n}$$

This sum is in general a function of α and β,; if, however, all the values x_n, y_n, and z_n are small in relation to λ; it becomes constant.

direction cosines of the incident primary X-radiation. We now need no further assumptions concerning the excitation process, except that it takes place in the same way for all atoms. We should note especially that for the further consideration of this process it is completely unimportant whether the oscillations in the atom are caused through oscillations of the same frequency in the primary radiation, or whether, when they have once been excited by the primary radiation, they then constitute free characteristic oscillations. We find throughout as the super-imposition effect of all elementary waves:

$$\sum \psi \; \frac{e^{-ik(r+x\alpha_0+y\beta_0+z\gamma_0)}}{r}. \tag{3}$$

We estimate this sum only for points whose distance is very great compared with the dimensions of the illuminated portion of the crystal, and also make use of the usual approximation in lattice-theory, in which instead of r in the numerator, the value R of the radius vector from the origin to the point of observation is used, and the function ψ is given the value corresponding to this direction (α, β, γ). For the value of r in the exponent, we use the approximate value:

$$r = R - (x\alpha + y\beta + z\gamma).$$

Taking account of equation (1), the summation (3) then becomes:

$$\psi(\alpha,\beta) \frac{e^{-ikR}}{R} \sum e^{ik[x(\alpha-\alpha_0)+y(\beta-\beta_0)+z(\gamma-\gamma_0)]}$$

$$= \psi(\alpha,\beta) \frac{e^{-ikR}}{R} \sum m \sum n \sum p \; e^{i(mA+nB+pC)} \tag{4}$$

where the abbreviations below are used:

$$\begin{aligned} A &= k[a_{1x}(\alpha-\alpha_0)+a_{1y}(\beta-\beta_0)+a_{1z}(\gamma-\gamma_0)] \\ B &= k[a_{2x}(\alpha-\alpha_0)+a_{2y}(\beta-\beta_0)+a_{2z}(\gamma-\gamma_0)] \\ C &= k[a_{3x}(\alpha-\alpha_0)+a_{3y}(\beta-\beta_0)+a_{3z}(\gamma-\gamma_0)]. \end{aligned} \tag{5}$$

If we think of the irradiated portion of the crystal as being bounded by planes parallel to the sides of an elementary parallelepiped, then the summation with respect to m from a number $-M$ to $+M$, with respect to n from $-N$ to $+N$, and with respect to p from $-P$ to $+P$ may be carried out. The position of the intensity maxima is independent of such assumptions. The intensity of the oscillation[5] then becomes:

$$\frac{|\psi(\alpha,\beta)|^2}{R^2} \frac{\sin^2 MA}{\sin^2 \frac{1}{2}A} \frac{\sin^2 NB}{\sin^2 \frac{1}{2}B} \frac{\sin^2 NC}{\sin^2 \frac{1}{2}C}. \tag{6}$$

Each of these sine-quotients attains its maximum when its denominator vanishes. The conditions for the maxima are thus:

$$A = 2h_1\pi, \text{ i.e. } a_{1x}\alpha + a_{2y}\beta + a_{1z}\gamma = h_1\lambda + a_{1x}\alpha_0 + a_{1y}\beta_0 + a_{1z}\gamma_0$$

$$B = 2h_2\pi, \text{ i.e. } a_{2x}\alpha + a_{2y}\beta + a_{2z}\gamma = h_2\lambda + a_{2x}\alpha_0 + a_{2y}\beta_0 + a_{2z}\gamma_0 \tag{7}$$

$$C = 2h_3\pi, \text{ i.e. } a_{3x}\alpha + a_{3y}\beta + a_{3z}\gamma = h_3\lambda + a_{3x}\alpha_0 + a_{3y}\beta_0 + a_{3z}\gamma_0.$$

The sums at the left-hand side of these equations are equal to the length of each edge multiplied by the cosine of the angle between it and the direction α, β, γ. Each of the equations (7) thus represents a family of circular cones whose axis corresponds with the direction of one of the edges a_1, a_2, and a_3.[6] It will, of course, only happen in exceptional cases that a direction will fulfil all three conditions simultaneously. It is this which causes the complication as compared with the simple or cross-lattice. Nevertheless a visible intensity maximum would be expected if the line of two cones of the first two families lies close to a cone of the third family. We shall consider this situation more closely for the case represented in Fig. 5, Plate II, in which a regular crystal is irradiated in the direction of one of the edges a_1, a_2, and a_3.

---

[5] Observation is in agreement with this.

[6] As in lattice-theory it is easily seen from the elementary geometrical construction of the path difference between two parallel lines for the rays from neighbouring particles, that this path difference on the cones here referred to is constant.

In this case the three edges have the same length a and are perpendicular to each other, so that we may allow the axes of co-ordinates to lie in the directions of these axes. Since then

$$a_{1y} = a_{1z} = a_{2x} = a_{2z} = a_{3x} = a_{3y} = 0, \quad a_{1x} = a_{2y} = a_{3z} = a,$$

while

$$\alpha_0 = 0, \quad \beta_0 = 0, \quad \gamma_0 = 1$$

then equation (7) becomes:

$$\alpha = h_1 \frac{\lambda}{a}, \quad \beta = h_2 \frac{\lambda}{a}, \quad 1 - \gamma = h_3 \frac{\lambda}{a}. \tag{8}$$

On a photographic plate perpendicular to the incident ray the curves α = constant and β = constant are hyperbolae with their centres at the point of incidence of the primary ray, and their axes perpendicular to each other. If both conditions in (8) were fulfilled, then the well-known cross-lattice spectrum would be observed, in which at each intersection point of two hyperbolae an intensity maximum occurs. If now we chose the curve γ = constant, whose mid-point also lies at the point of incidence of the primary ray, the whole of the curve will not be found on the plate, but only a series of individual points from the cross-lattice spectrum which lies sufficiently near to one of the two hyperbolae. This is in·fact the impression created by the figure. The curves

$$1 - \gamma = h_3 \frac{\lambda}{a}, \tag{9}$$

which are due to the periodicity in the direction of the ray, have an analogy in optics, known since the days of Newton, in the so-called Quetelet rings.[7] These are formed in the presence of dust on a smooth glass plate, strongly reflecting on the back, in the Fraunhöfer diffraction picture. What occurs is that the incident and reflected light waves at the same dust particle undergo

---

[7] Cf. A. Winkelmann, *Handbuch der Physik*, VI (Leipzig, 1906), p. 1083.

interference with one another. For perpendicular incidence the maxima lie on the curve given by equation (9), but are much flatter than in the case we are considering, since only two waves undergo interference. If the dust particles could be arranged in a regular cross-lattice, then the analogy with the interference at crystals would undoubtedly be much closer.

It must be observed that in the space lattice under consideration the division into elementary parallelepipeds is not a unique one, but can be undertaken in an infinite number of ways. In the case of the regular space lattice, for example, instead of breaking this down into cubes, it is possible to use parallelepipeds of which one side is the diagonal of a cube surface, so that the direction has a twofold axis of symmetry. The intensity maxima in such a case would correspond to a broken conical section on axes appropriate to each type of sub-division of the crystal. In actual fact, Fig. 12, Plate V, confirms that the spots are arranged on a curve on a twofold axis, when the regular crystal is irradiated in this direction and the photographic plate is placed perpendicular to it. Even in this complicated figure the impression is always given that the intensity maxima can be represented by broken conical sections. As far as we can at present see, the theory is well applicable to the fourfold symmetrical postulate if we assume the presence of several wavelengths from $0.038\ a$ to $0.15\ a$. Since a for zinc sulphide is 3.38×10^{-8} cm, the wavelengths would then lie within the range 1.3×10^{-9} to 5.2×10^{-9}.

In spite of this agreement it is impossible to deny that our theory requires comprehensive improvement. The thermal motion of the molecules distorts these even at room temperature by an appreciable fraction of the lattice-constant, and consequently by many times the wavelength, a fact requiring particular attention. It would, moreover, be premature to draw conclusions from equation (6) as to the sharpness of the interference maxima.

The fact that of all crystals so far investigated, with the exception of diamond, the intensity maxima are limited to a sharp angle with respect to the direction of the primary ray, instead of emerging in all directions as one would have at first

D*

expected from equation (6), may presumably be referred to the direction-function ψ; it is, however, also conceivable that the thermal movement must be invoked in explaining this phenomenon.

General Conclusions

We discuss finally, without any reference to the formula, the question as to how far these experiments indicate a wave-nature for X-radiation.

Radiation emerging from a crystal possesses a wave-character as evidenced by the sharpness of the intensity maxima, which is easily understood as an interference phenomenon, but can scarcely be understood on the basis of a corpuscular theory. The high penetrating power of the radiation points in the same direction, since of all known corpuscular rays, only the fastest β-rays are capable of this. It may well be possible also to demonstrate the wave character of the primary radiation. Let us imagine that the atoms of the crystal in the case of Fig. 5, Plate II, are excited by means of a corpuscular radiation. (The light-quantum structure of X-radiation assumed by many investigators may here be included under the designation corpuscular radiation.) Then only those atoms which were struck by the same corpuscle would be brought into coherent oscillations, which would only apply to the series of atoms parallel to the z-direction. Atoms at a definite distance from each other in the x- or y-direction would be excited by different corpuscles, and a definite phase difference between their oscillations could therefore not arise. Because of this, in the intensity expression (6) only one sine-quotient would remain; we would have only one condition for an intensity maximum, which, as symmetry considerations also indicate, would be fulfilled along circles around the point of impact of the primary ray. The broken character of these circles which is actually observed would be completely incomprehensible. Moreover, the primary radiation and that emerging from a crystal are in all appearances so similar that the wave-nature of the latter may be regarded as definite evidence of the wave-nature of the former also. There is certainly one distinction: the radiation emergent from a crystal has a definite spectroscopic homogeneity, that is, a certain periodicity. The primary radiation, on the other hand, is, according to Sommerfeld's theory, in so far as it consists of Bremsstrahlung, made up of non-periodic impulse waves. The experiments which are now to be discussed are in any case compatible with this assumption. We will for the present leave undecided whether the periodic radiation is formed in the crystal through fluorescence or whether it is already present in the primary radiation itself, together with the pulses, and is simply separated by the crystal.[8] There is ground for hoping that further investigation will soon make it possible to decide this question.

Experimental Part—W. Friedrich and P. Knipping

For an experimental test of the views set out in the previous part of this article, after a few preliminary experiments with a provisional apparatus, the following definitive experimental arrangement was employed, as represented schematically in Fig. 1. From the anticathode A of a Röntgen tube, a narrow beam of about 1 mm diameter is selected from the emitted X-rays by means of the shutters B_1 to B_4. This beam passes through the crystal Kr, which is located in a Goniometer G. Around the crystal photographic plates P were located at various distances and in various directions, so that the intensity distribution of the secondary radiation emitted from the crystal could be recorded on them. Adequate protection against undesired radiation was provided for in the apparatus by means of a large lead screen S, and the lead case K. The size of the important parts of the apparatus can be gathered from Fig. 1 which is drawn to a scale of (approximately) 1:10.

The positioning of the parts of the complete apparatus was performed by an optical method. A cathetometer, whose telescope

---

[8]See also p. 107.

possessed a cross-wire, had been permanently fixed in a rigid position.   Along the optical axis of the telescope were then arranged in series the focus of the anticathode, the shutters and the axis of the goniometer.   In this way the very troublesome task of arranging the apparatus by means of X-rays could be replaced by the much more convenient optical method.   Control-led experiments with X-rays showed that this adjustment by the

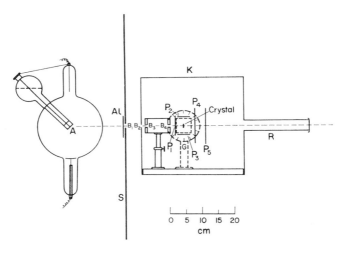

FIG. 1. Distance from anticathode to crystal 350 mm;   distance from crystal-P_1, P_2 or P_3 25 mm;   distance from crystal-P_4 35 mm; distance from crystal-P_5 70 mm.

optical method was completely satisfactory.   The shutters B_1 to B_3 were mainly employed in screening off the secondary radiation from the tube walls, while the actual limitation of the X-ray beam falling upon the crystal was performed by means of shutter B_4. This had in general a diameter of 0·75 mm, and was bored in a lead plate 10 mm thick.   It could be adjusted by means of three locating screws (which are not shown) in such a way that the axis of the hole corresponded exactly with that of the telescope and of the X-ray beam.   By this means it could be ensured that an X-ray beam of circular cross-section fell upon the crystal, and also that the amount of secondary radiation emitted from the wall of this shutter B_4 was reduced to a minimum, which was necessary in order to suppress background on the photographic plates. So as to be able to check the alignment during operation, we also photographed the primary beam after allowing it to pass out of the protective box by means of a long supplementary tube R, a special casette being located at the end of this tube (not shown in the diagram). The tube R served the purpose of avoiding as far as possible the secondary radiation arising through the impact of the primary rays on the rear wall of the protective case.

After these adjustments, which were checked before each experiment, the goniometer axis was arranged perpendicular to the radiation path by the usual method. In the same way, the various plate holders were so adjusted that the primary radiation was oriented either perpendicular or parallel to the film or plate used. When the apparatus was arranged to this extent, the crystal intended for irradiation, which was fixed to the goniometer table by means of a trace of adhesive wax, was put in place, and then by means of the telescope already mentioned, located by the well-known method using a "signal". This very important adjustment, the importance of which will be discussed later, could be carried out to within an accuracy of 1 minute of arc. As sensitive material for the photographic plates, after we had rejected a number of other types as unsuitable, we used Schleussner–Röntgen films, which gave the best results when developed with Rodinal (1:15).

Approximate calculations on the basis of previous experience of secondary radiation showed that very considerable exposure times were necessary. The time of illumination with a current of from 2 to 10 milli-amperes (according to the hardness of the tube between 6 and 12 Wehnelt), varied from 1 to 20 hours.[9] The X-ray tubes used were partly tubes from Gundelach, and partly

---

[9] Two hardness values were kindly placed at our disposal by the firm of Reiniger, Gebbert and Schall.

tubes with water-cooling from Müller, which were operated from a 50 cm Klingelfuss inductor. Either a Wehnelt or a mechanical interrupter was employed. Before the X-ray tube suitable valves were used to avoid sparking. The exposure was carried out intermittently to avoid over-heating of the tube. For the experiments carried out up to now a hardness of 8–10 Wehnelt has been found to be suitable.

Preliminary experiments with the provisional apparatus[10]

Since we supposed initially that we were concerned with fluorescence radiation it was necessary to use a crystal containing a metal of considerable atomic weight, in order to obtain secondary radiation as intense and homogeneous as possible, as this seemed to be most suitable for the experiments. According to Barkla metals of atomic weight lying between 50 and 100 would be most suitable. Since we had initially no good crystals containing such metals, we used a reasonably good copper sulphate crystal for the preliminary investigations. This was placed more or less in an arbitrary position in the apparatus, and the X-rays fell approximately vertically on a 110 face. At a distance of 40 mm from this were located two photographic plates corresponding to P_2 and P_4 in Fig. 1. After the exposure the upper plate was weakly but uniformly blackened, while P_4 showed, outside the point of impact of the primary ray, a series of apparently regularly arranged spots.

In order to make certain that these spots were due to the crystal structure of the copper sulphate, copper sulphate crystals were roughly powdered, placed in a small paper carton and used for a repetition of the previous experiment under otherwise identical conditions. Plate P_2 showed no alteration, but the large regularly arranged spots on P_4 had vanished, while instead of these there appeared numerous small irregularly placed points around the point of impact of the primary radiation (see Plate I, Fig. 2).

---

[10]This apparatus was similar in principle to that used later, but simpler in construction and without means of accurate adjustment. The shutter openings amounted to 3 mm.

The expectation that with sufficiently finely powdered material the remainder of the spots would vanish was confirmed in the next experiment. A third experiment with crystals taken from this under otherwise identical conditions showed, as was to be expected, that the plates were unblackened right up to the point of impact of the primary radiation.

Two further exposures were then made with the copper sulphate crystals first referred to. In the first of these, the crystal, while preserving its orientation, was shifted parallel to itself, so that another portion of the crystal was irradiated; in the second of them the crystal was inclined a few degrees towards the tube. The first exposure was identical with Fig. 1, Plate I, so that the phenomenon is independent of the portion of crystal which is irradiated. The other exposure shows that the position of the secondary spots depends upon the orientation of the crystal with respect to the angle of incidence of the primary radiation.

The experiments described above showed that the ideas developed previously are experimentally verifiable, so that it seemed appropriate to build the improved apparatus already described, though in the meanwhile further experiments were carried out using zinc blende, rock salt and lead glance (cleavage fragments). These gave quite similar results to those obtained with copper sulphate. We will not, however, give a description of these experiments, since it appeared that a more accurate orientation of the crystals was necessary in order to get a closer insight into the phenomenon.

Experiments with more accurate orientation of the crystals with respect to the primary radiation

The first experiment with the definitive apparatus consisted in the exposure depicted in Fig. 1, Plate I, in which all the plates shown in Fig. 1 were in position.[11] The orientation of the copper sulphate was as far as possible the same as previously. Plates P_2 and P_3 showed the same degree of blackening; P_1 showed only

---

[11] The shutter opening B_4 in this experiment was 1·5 mm.

the point of impact of the primary radiation. P_4 and P_5 showed the same pattern as that obtained in exposure 1, except that, because of the smaller shutter opening, the secondary spots were compressed together (see Plate I, Figs. 3 and 4). It is noteworthy that the ratio of the distances crystal-P_4 to crystal-P_5 is the same as the ratio of the sizes of the patterns on P_4 and P_5, from which it appears that the radiation proceeds in straight lines from the crystal. It should also be noted that the size of the individual secondary spots, in spite of the greater distance of plate P_5 from the crystal, has remained unchanged. This may well be an indication that the secondary radiation corresponding to each individual spot leaves the crystal in the form of a parallel pencil.

It would be expected that the phenomenon with crystals of the cubic system would be clearer and easier to understand than in the case of the triclinic copper sulphate, since the corresponding spatial lattice in the case of such crystals is the simplest possible. A suitable substance for investigating this seemed to be zinc blende, with which, as we have already indicated, experiments have been made and appreciable intensity of secondary radiation obtained. We obtained from Steeg and Reuter in Homburg a plain parallel plate cut from a good crystal parallel to a cubic face (100) (perpendicular to a main crystallographic axis) of size $10 \times 10 \times 0.5$ mm. This plate was accurately oriented in the way described above so that the primary radiation struck the crystal perpendicular to the cube face. The result of such an experiment is shown in Fig. 5,[12] Plate II. The position of the secondary spots is completely symmetrical in relation to the point of impact of the primary radiation. It is possible to see in the figure two pairs of planes of symmetry arranged perpendicular to each other. If one takes any of the spots from the figure and if this does not lie on one of the planes of symmetry, it is possible to bring it into the same position as seven corresponding points by means of reflexion and rotation through 90°. If a point falls upon a plane

---

[12]Exposure 9 in Plate IV was made with a provisionally located film 10 mm away from the crystal. In Fig. 10, Plate IV, the film was at twice this distance from the crystal, as compared with Fig. 5, Plate II.

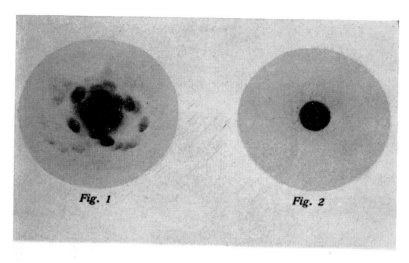

Fig. 1

Fig. 2

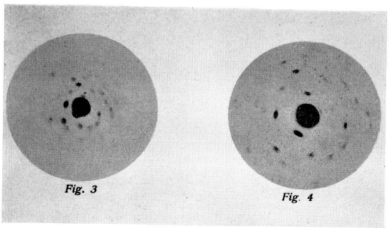

Fig. 3

Fig. 4

PLATE I.

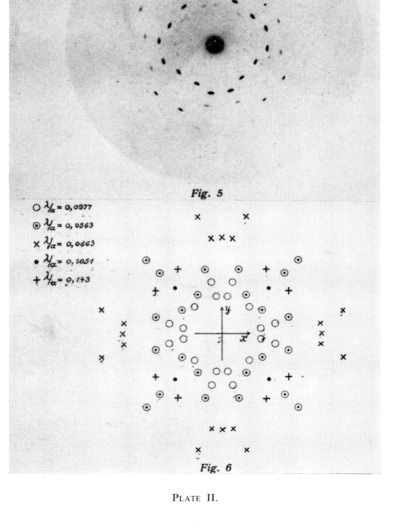

Fig. 5

○ $\lambda/_a = 0,0377$
⊙ $\lambda/_\alpha = 0,0563$
× $\lambda/_\alpha = 0,0663$
• $\lambda/_\alpha = 0,1051$
+ $\lambda/_\alpha = 0,143$

Fig. 6

PLATE II.

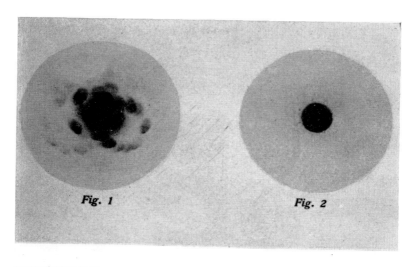

Fig. 1

Fig. 2

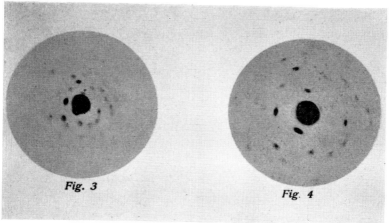

Fig. 3

Fig. 4

PLATE I.

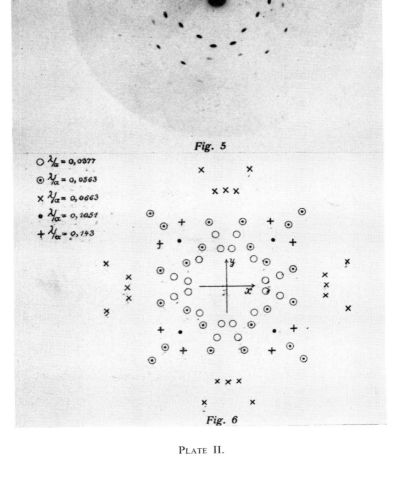

Fig. 5

$\lambda/\alpha = 0,0377$
$\lambda/\alpha = 0,0563$
$\lambda/\alpha = 0,0663$
$\lambda/\alpha = 0,1051$
$\lambda/\alpha = 0,143$

Fig. 6

PLATE II.

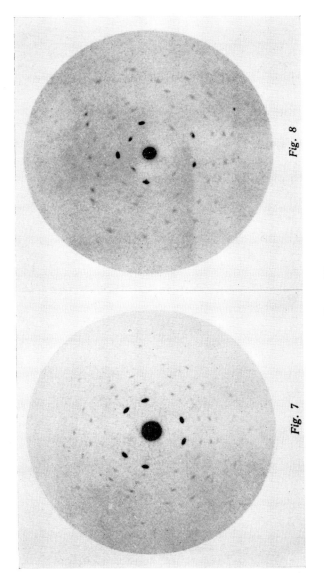

Fig. 8

Fig. 7

PLATE III.

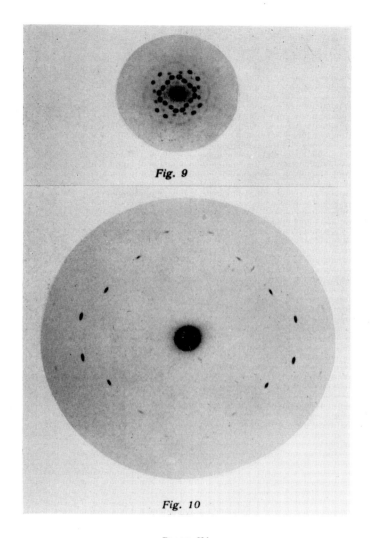

Fig. 9

Fig. 10

PLATE IV.

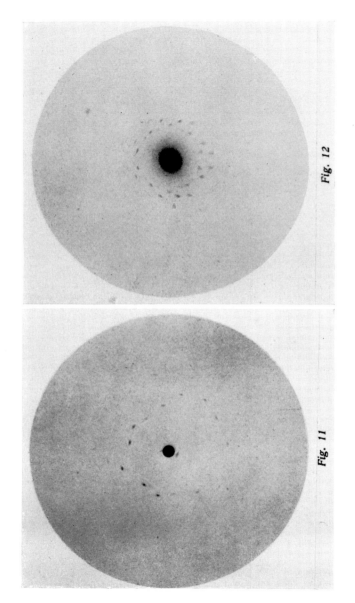

Fig. 12

Fig. 11

of symmetry, this can naturally only be brought into correspond-ence with three other points. This corresponds to holohedral symmetry of the cubic system, in spite of the fact that zinc blende corresponds to a hemihedral class. This fact that a completely fourfold symmetry is present on the plate is certainly one of the most beautiful demonstrations of the space-lattice of the crystal, and of the fact that no other property other than the space-lattice is involved. Since the space-lattice always shows holo-hedral symmetry translation of space-lattices of different nature with respect to each other, as must be assumed for the explanation of the hemihedra, would according both to experiment and our theory be without effect. This requirement of symmetry has subsequently and without exception been vindicated by experi-ment. On one occasion we allowed the primary radiation to impinge vertically on an octahedral plane (111), and then upon a rhombododecahedral plane (110). Figure 7, Plate III, gives the threefold symmetry corresponding to the threefold axis along which the crystal was irradiated; Fig. 12, Plate V, gives similarly a twofold symmetry corresponding to the twofold axis. As to the latter, the information given below should be used for com-parison.

If we rotate the crystal round a primary ray, we must conclude from what has already been said that the image on the plate would be rotated correspondingly. In actual fact, experiments completely confirm this conclusion.

It was necessary to show that an accurate orientation of the crytal is necessary in order to obtain identical pictures on repetition of an experiment. We allowed the primary radiation to move through 3° with respect to a fourfold axis in such a way that the plane through the ray and the axis was a plane of sym-metry. The result of this experiment is given in Fig. 11, Plate V. As we can see, the fourfold symmetry has vanished, although the majority of the spots in Fig. 5, Plate II, can still be recognized. One plane of symmetry appears, however, here also. In a similar way Fig. 8, Plate III, arises by a movement of the crystal by 3° with respect to Fig. 7, Plate III.

If we rotated a plate ground perpendicularly to the fourfold axis, in such a way that the primary ray proceeded parallel to a threefold axis, a figure was obtained which was distinguished from Fig. 7, Plate III, not in the position of the spots, but only in their size. According to the symmetry, also, equivalent spots showed the same difference in size. This showed that the limit-ation of the portion of crystal through which the radiation passes

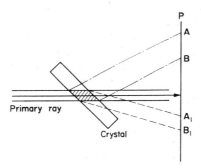

FIG. 2. Primary ray.

is without effect on the position of the intensity maxima, while the difference in the size of the spots corresponds to the difference of the variously directed projections of the irradiated crystal portion on the photographic plate (see Fig. 2). The shaded area represents the irradiated portion of the crystal. AB and A_1B_1 are the projections of this portion on the plate P and give the size of the spots. If we rotated the same crystal plate in such a way that the primary ray now struck the crystal in the direction of a twofold axis of symmetry, Fig. 12, Plate V, was produced, which differed only in the unequal size of corresponding dots from an exposure obtained with a cleavage fragment obtained perpendicular to a twofold axis.

In addition to the experiments already described we have carried out further work with zinc blende, copper crystals, rock salt, and a diamond plate. (The diamond plate is a very generous gift to the Physical Institute of the University by the firm of Ginsberg in Hanau.) We are at present engaged on the photograms, and though the experiments are not as yet complete, we hope to be able to report on them shortly. We wish, however, to refer to two experimental results from this series which seem to us to be significant.

In the first place, we have found in the case of rock salt that the intensity of the secondary spots depends upon the thickness of the irradiated layer. The blackening obtained for the same time of exposure, the same tube-hardness and otherwise equivalent conditions, for a 15 mm layer of the crystal was considerably greater than that with a 1·5 mm layer. We intend to continue these experiments with other materials.

In the second place we should like to devote a short space to an experiment with diamond. According to Barkla carbon shows no characteristic radiation so far found. It would therefore be expected that if the spots for the other crystals are due to characteristic radiation from the crystal, no such dots would be shown by diamond, and only general blackening due to scattered radiation would be found. Contrary to this expectation, not only do the rear plates P_4 and P_5 show similar phenomena to those obtained in earlier exposures, but also plates P_1, P_2, and P_3 give clearly distinguished spots.[13] Whether this remarkable fact is connected with the small atomic volume of diamond in comparison with that of the other crystals, or with the anomalous behaviour of this substance with respect to thermal oscillations (revealed in the specific heat), we must for the present regard as uncertain.[14] Decisive experiments on this question are at present being considered.

In addition to the experiments so far described, which were intended to investigate the connection between the position of

---

[13]Even when the exposure lasts for a period of days we have, in the case of other crystals, been unable to obtain anything other than general blackening on plates P_1, P_2, and P_3.

[14]See Theoretical portion.

the secondary spots and the lattice structure, we have also investigated the nature of the radiation producing the spots. That we are here concerned with secondary X-radiation is made probable by the fact that the radiation is able to penetrate through considerable thicknesses of metals, such as the sheet steel plate holder. Provisional hardness measurements have been made in the following way using a zinc blende crystal ground perpendicular to the fourfold axis. Since we do not know the law governing blackening due to X-radiation, the intensity of the ray has been taken as proportional to the blackening produced, and the blackening has been measured by means of a Hartmann microphotometer. The crystal symmetry requires that eight spots in a regular circle should produce the same blackening, which in fact proves to be the case. In front of plate P_5 we introduced an aluminium sheet 3 mm thick, which covered two opposite quarters of the plate, leaving the other two free. The primary pencil passed through a central opening in order to avoid producing secondary radiation from the aluminium sheet. Measurements of blackening, shown in Table I, reveal that the absorption coefficient of the rays is approximately 3.84 cm^{-1} for aluminium. Column 1 gives the diameter of the circle on which the corresponding spots lie. In column 2 we give the measured blackening of the spots in the absence of the aluminium sheet, while in column 3 we give the blackening after passage of the rays through the aluminium sheet. Column 4 gives values for the background blackening of the plates, while the last column gives the calculated absorption coefficient k for aluminium calculated from the known absorption law. A hardness measurement for diamond carried out according to the same method gave a hardness of secondary radiation of about the same magnitude. The hardness measurements by the photographic method are, however, not sufficiently accurate to permit of definite conclusions, since the difference in hardness found lies outside the observational error. By means of an electrical method, which is free from the uncertainties of the photographic method, particularly faults in the plates, these experiments are to be continued, and the results will be published later.

Barkla finds for the penetrating power of the characteristic radiation of zinc an appreciably lower value, corresponding to an appreciably larger value of k. We do not, however, regard as absolutely necessary the conclusion that we are not dealing with characteristic radiation. The following facts indicate fluorescence radiation which, however, may possibly be already present in the primary radiation: the same hardness of the rays for zinc sulphide and diamond suggest that this may well be true. During the time of exposure the hardness of the primary radiation diminished,

TABLE I

ϕ	Without Al	With Al	Background	k
50 mm	1·78	0·88	0·45	$3·53 \text{ cm}^{-1}$
57·5 mm	2·02	0·79	0·45	$4·71 \text{ cm}^{-1}$
74 mm	0·69	0·53	0·45	$3·23 \text{ cm}^{-1}$
86 mm	0·84	0·55	0·45	$3·89 \text{ cm}^{-1}$

especially in the case of older tubes, which needed frequent regeneration, to a considerable extent (from 6 to 12 Wehnelt). All the same, the secondary spots remained, as the photograms showed, sharply defined and undistorted. A comparison of the hardness of the rays lying on different rings indicates that these consist of radiation with different penetrating powers. It would therefore not be surprising to find that we are here dealing with characteristic radiation of the crystal: since then the crystal is constructed of zinc and sulphur atoms, both the zinc and sulphur atoms would be able to emit radiation, which are known from previous experiments to be of different hardness. If the characteristic radiation arises from the anticathode itself, according to Barkla we should expect more radiation in the case of platinum.

In conclusion we should like to express our hearty thanks to Professor A. Sommerfeld, in whose Institute this work has been carried out, both for his ready provision of the apparatus and for his constant interest and valuable counsel. We must also thank Geheimrat Röntgen and Geheimrat v. Groth for the provision of valuable crystals and apparatus, as well as for their keen interest in the work.

University of Munich, Institute of Theoretical Physics.

2. The Diffraction of Short Electromagnetic Waves by a Crystal*

W. L. BRAGG

HERREN FRIEDRICH, Knipping, and Laue have lately published a paper entitled 'Interference Phenomena with Röntgen Rays[1],' the experiments which form the subject of the paper being carried out in the following way. A very narrow pencil of rays from an X-ray bulb is isolated by a series of lead screens pierced with fine holes. In the path of this beam is set a small slip of crystal, and a photographic plate is placed a few centimetres behind the crystal at right angles to the beam. When the plate is developed, there appears on it, as well as the intense spot caused by the undeviated X-rays, a series of fainter spots forming an intricate geometrical pattern. By moving the photographic plate backwards or forwards it can be seen that these spots are formed by rectilinear pencils spreading in all directions from the crystal, some of them making an angle of over 45° with the direction of the incident radiation.

When the crystal is a specimen of cubical zinc blende, and one of its three principal cubic axes is set parallel to the incident beam, the pattern of spots is symmetrical about the two remaining axes. This pattern is shown in Plate I. Laue's theory of the formation of this pattern is as follows. He considers the

Proceedings of the Cambridge Philosophical Society, **17**, 43–57 (1913). Communicated by Professor Sir J. J. Thomson. Read 11 November 1912.

[1] *Sitzungsberichte der Königlich Bayerischen Akademie der Wissenschaften*, June 1912.

molecules of the crystal to form a three-dimensional grating, each molecule being capable of emitting secondary vibrations when struck by incident electromagnetic waves from the X-ray bulb. He places the molecules in the simplest possible of the three cubical point systems, that is, molecules arranged in space in a pattern whose element is a little cube of side 'a', with a molecule at each corner. He takes coordinate axes whose origin is at a point in the crystal and which are parallel to the sides of the cubes. The incident waves are propagated in a direction parallel to the z axis, and on account of the narrowness of the beam the wave surfaces may be taken to be parallel to the xy plane. The spots are considered to be interference maxima of the waves scattered by the orderly arrangement of molecules in the crystal. In order to get an interference maximum in the direction whose cosines are α, β, γ, for incident radiation of wavelength λ, the following equations must be satisfied

$$a\alpha = h_1\lambda, \quad a\beta = h_2\lambda, \quad a(1-\gamma) = h_3\lambda \tag{1}$$

where h_1, h_2, h_3 are integers.

These equations express the condition that the secondary waves of wavelength λ from a molecule, considered for simplicity as being at the origin of coordinates, should be in phase with those from its neighbours along the three axes, and that therefore the secondary waves from all the molecules in the crystal must be in phase in the direction whose cosines are α, β, γ.

The distance of the crystal from the photographic plate in the experiment was 3·56 cm. The pencil of X-rays on striking the crystal had for cross-section a circle of diameter about a millimetre, and the dimensions of the spots are of the same order. The plate of crystal was only ·5 millimetre thick. It is thus easy to calculate with considerable accuracy from the position of a spot on the photographic plate the direction cosines of the pencil to which it corresponds, since the pencils of rays may be all taken as coming from the centre of the crystal. Laue found, on doing this for each spot, that as a matter of fact the values for

α, β, $1-\gamma$ so obtained were in the numerical ratio of three small integers h_1, h_2, h_3 as they should be by equations (1).

For instance, a spot appears on the photographic plate whose coordinates referred to the x and y axes are

$$x = \cdot 28 \text{ cm.}, \quad y = 1\cdot 42 \text{ cm.}$$

The distance of the crystal from the photographic plate, 3·56 cm., gives z.

Thus since $\qquad \alpha : \beta : \gamma :: x : y : z$

$$\frac{\alpha}{\cdot 28} = \frac{\beta}{1\cdot 42} = \frac{\gamma}{3\cdot 56} = \frac{1}{\sqrt{(\cdot 28)^2 + (1\cdot 42)^2 + (3\cdot 56)^2}} = \frac{1}{3\cdot 83}$$

Thus $\qquad\qquad \dfrac{\alpha}{\cdot 28} = \dfrac{\beta}{1\cdot 42} = \dfrac{1-\gamma}{\cdot 27},$

or $\qquad\qquad\qquad \alpha : \beta : 1-\gamma :: 1 : 5 : 1.$

Laue considers some thirteen of the most intense spots in the pattern. Owing to the high symmetry of the figure, the whole pattern is a repetition of that part of it contained in an octant. Thus these thirteen represent a very large proportion of all the spots in the figure. For these spots he obtains corresponding integers h_1, h_2, h_3 which are always small, the greatest being the number 10. But even if one confines oneself to integers less than 10, there are a great many combinations of h_1, h_2, h_3 which might give spots on the photographic plate which are in fact not there, and there is no obvious difference between the numbers h_1, h_2, h_3 which correspond to actual spots, and those which are not represented.

To explain this Laue assumes that only a few definite wavelengths are present in the incident radiation, and that equations (1) are merely approximately satisfied.

Considering equations (1) it is clear that when h_1, h_2, h_3 are fixed λ/a can only have one value. However if h_1, h_2, h_3 are multiplied by an integral factor p, equations (1) can still be satisfied, but now by a wavelength λ/p. By adjusting the numbers h_1, h_2, h_3 in this way, Laue accounts for all the spots considered by means of five different wavelengths in the incident radiation. They are

$$\lambda = \cdot 0377a$$
$$\lambda = \cdot 0563a$$
$$\lambda = \cdot 0663a$$
$$\lambda = \cdot 1051a$$
$$\lambda = \cdot 143a.$$

For instance, in the example given above, where it was found that

$$\alpha : \beta : 1 - \gamma :: 1 : 5 : 1$$

these numbers are multiplied by 2, becoming 2:10:2. Then they can be assigned to a wavelength

$$\lambda / a = \cdot 037,$$

approximately equal to the first of those given above.

However, this explanation seems unsatisfactory. Several sets of numbers h_1, h_2, h_3 can be found giving values of λ/a approximating very closely to the five values above and yet no spot in the figure corresponds to these numbers. I think it is possible to explain the formation of the interference pattern without assuming that the incident radiation consists of merely a small number of wavelengths. The explanation which I propose, on the contrary, assumes the existence of a continuous spectrum over a wide range in the incident radiation, and the action of the crystal as a diffraction grating will be considered from a different point of view which leads to some simplification.

Regard the incident light as being composed of a number of independent pulses, much as Schuster does in his treatment of the action of an ordinary line grating. When a pulse falls on a plane it is reflected. It if falls on a number of particles scattered over a plane which are capable of acting as centres of disturbance when struck by the incident pulse, the secondary waves from them will build up a wave front, exactly as if part of the pulse had been reflected from the plane, as in Huyghens' construction for a reflected wave.

The atoms composing the crystal may be arranged in a great many ways in systems of parallel planes, the simplest being the cleavage planes of the crystal. I propose to regard each interference maximum as due to the reflection of the pulses in the incident beam in one of these systems. Consider the crystal as divided up in this way into a set of parallel planes. A minute fraction of the energy of a pulse traversing the crystal will be reflected from each plane in succession, and the corresponding interference maximum will be produced by a train of reflected pulses. The pulses in the train follow each other at intervals of $2d \cos \theta$, where θ is the angle of incidence of the primary rays on the plane, d is the shortest distance between successive identical planes in the crystal. Considered thus, the crystal actually 'manufactures' light of definite wavelengths, much as, according to Schuster, a diffraction grating does. The difference in this case lies in the extremely short length of the waves. Each incident pulse produces a train of pulses and this train is resolvable into a series of wavelengths λ, $\lambda/2$, $\lambda/3$, $\lambda/4$ etc. where $\lambda = 2d \cos \theta$.

Though to regard the incident radiation as a series of pulses is equivalent to assuming that all wavelengths are present in its spectrum, it is probable that the energy of the spectrum will be greater for certain wavelengths than for others. If the curve representing the distribution of energy in the spectrum rises to a maximum for a definite λ and falls off on either side, the pulses may be supposed to have a certain average 'breadth' of the order of this wavelength. Thus it is to be expected that the intensity of the spot produced by a train of waves from a set of planes in the crystal will depend on the value of the wavelength, viz. $2d \cos \theta$. When $2d \cos \theta$ is too small the successive pulses in the train are so close that they begin to neutralize each other and when again $2d \cos \theta$ is too large the pulses follow each other at large intervals and the train contains little energy. Thus the intensity of a spot depends on the energy in the spectrum of the incident radiation characteristic of the corresponding wavelength.

Another factor may influence the intensity of the spots. Consider a beam of unit cross-section falling on the crystal. The strength of a pulse reflected from a single plane will depend on the number of atoms in that plane which conspire in reflecting the beam. When two sets of planes are compared which produce trains of equal wavelength it is to be expected that if in one set of planes twice as many atoms reflect the beam as in the other set, the corresponding spot will be more intense. In what follows I have assumed that it is reasonable to compare sets of planes in which the same number of atoms on a plane are traversed by unit cross-section of the incident beam, and it is for this reason that I have chosen the somewhat arbitrary parameters by which the planes will be defined. They lead to an easy comparison of the effective density of atoms in the planes. The effective density is the number of atoms per unit area when the plane with the atoms on it is projected on the xy axis, perpendicular to the incident light.

Laue considers that the molecules of zinc-blende are arranged at the corners of cubes, this being the simplest of the cubical point systems. According to the theory of Pope and Barlow this is not the most probable arrangement. For an assemblage of spheres of equal volume to be in closest packing, in an arrangement exhibiting cubic symmetry, the atoms must be arranged in such a way that the element of the pattern is a cube with an atom at each corner and one at the centre of each cube face. With regard to the crystal of zinc-blende under consideration zinc and sulphur being both divalent have equal valency volumes and their arrangement is probably of this kind. It will be assumed for the present that the zinc and sulphur atoms are identical as regards their power of emitting secondary waves.

Take the origin of coordinates at the centre of any atom, the axes being parallel to the cubical axes of the crystal. The distance between successive atoms of the crystal along the axes is taken for convenience to be $2a$.

All atoms in the xz plane will have coordinates

$$pa \quad o \quad qa$$

where p and q are integers and $p+q$ is even. See Fig. 1 in text.

The same holds for atoms in the yz plane. Therefore any

reflecting plane may be defined by saying that it passes through the origin, and the centres of atoms

$$
\begin{array}{ccc}
pa & o & qa \\
o & ra & sa
\end{array}
$$

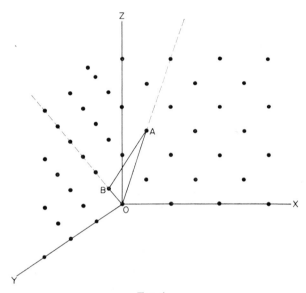

Fig. 1.

For instance, the plane on which the triangle OAB lies passes through the origin and

$$
\begin{array}{ccc}
a & o & 3a \\
a & o & a
\end{array}
$$

The planes can now be classified by the corresponding values of p, q, r, s as parameters.

The direction cosines of a plane $p\ q\ r\ s$ will be

$$
\frac{rq}{\sqrt{p^2s^2+q^2r^2+p^2r^2}}, \ \frac{ps}{\sqrt{p^2s^2+q^2r^2+p^2r^2}}, \ \frac{-pr}{\sqrt{p^2s^2+q^2r^2+p^2r^2}}.
$$

If these are called l, m, n the direction cosines of the reflected beam are

$$2ln, \quad 2mn, \quad 2n^2 - 1,$$

and the position of the interference maximum on the photographic plate can be found in terms of these quantities.

The corresponding wavelength is $2d \cos \theta$ where d is the perpendicular distance between successive planes. Now θ is the angle of incidence, therefore $\cos \theta = n$ above. It is easier to find the intercepts which successive planes cut off on the z axis, than their perpendicular distance apart. Calling these intercepts l, then

$$\lambda = 2d \cos \theta = 2 \cdot l \cos \theta \cdot \cos \theta = 2ln^2.$$

Consider the atoms as arranged in vertical rows parallel to the z axis in the figure. A plane for which $p = 1$ and $r = 1$ passes through one atom in every one of these vertical rows. Therefore the next plane to it passes through a set of atoms all $2a$ above the corresponding atoms in the first plane. Thus for this set of planes, $l = 2a$ and the wavelength $\lambda = 4an^2$. The effective density of atoms on such a set of planes is the greatest possible.

If $p = 1$, $r = 2$, each plane now passes through atoms in one half of the vertical rows. For instance, the plane through the origin contains no atoms in those vertical rows for which r is odd. The successive planes must cut the z axis at intervals $2a/2$, since the effective density of atoms in each is half as great as before and the whole number of atoms in unit volume of the crystal remains constant. Similarly if $p = 1$, $r = 3l = 2a/3$ and so forth.

In the general case $l = \dfrac{2a}{\text{L.C.M. of } p \text{ and } r}$.

In the tables given below planes with the same effective density of atoms on them, and therefore the same values of l, are grouped together.

The position of the spot reflected by each system of planes considered has been calculated, also the wavelength of the reflected train expressed for convenience in the form a/λ, and when in the photograph a spot is visible in the position calculated, its intensity is denoted by star according to an arbitrary scale.

$$\textbf{\Large *} \quad * \quad * \quad + \quad \bullet$$

When no spot appears in the calculated position, I have put 'invisible' opposite that plane.

TABLE I. PLANES FOR WHICH $p=1$, $r=1$, $l=2a$, $\lambda=4an^2$

p	q	r	s	a/λ	Intensity	h_1	h_2	h_3
1	1	1	3	2·8	*	1	3	1
1	1	1	5	6·8	✸	1	5	1
1	1	1	7	12·8	*	1	7	1
1	1	1	9	20·8	Invisible	1	9	1
1	3	1	1	2·8	*	3	1	1
1	3	1	3	4·8	*	3	3	1
1	3	1	5	8·8	✸	3	5	1
1	3	1	7	14·8	+	3	7	1
1	3	1	9	22·8	Invisible	3	9	1
1	5	1	1	6·8	✸	5	1	1
1	5	1	3	8·8	✸	5	3	1
1	5	1	5	12·8	*	5	5	1
1	5	1	7	18·8	Invisible	5	7	1
1	7	1	1	12·8	*	7	1	1
1	7	1	3	14·8	+	7	3	1
1	7	1	5	18·8	Invisible	7	5	1
1	9	1	1	20·8	Invisible	9	1	1

Range of values of a/λ, all possible up to 15.

There is no need to go any further than the set for which $l = 2a/4$, to obtain all the spots in the photograph. Indeed only one spot is given by this last set.

Only one spot on the plate is to be assigned to planes of this class. It is curious that the value of a/λ corresponding to this spot should be as great as 11·2. It is noticeable in the photograph that all spots at any distance from the centre of the pattern tend to become very faint, and the values of p, q, r, s which do give a spot in Table IV are the only ones to be found giving a spot at all near the centre. In the first three tables the parameters corresponding to a value of a/λ between 6 and 9 are represented by the most intense spots.

Every spot in the photograph is accounted for in the following Tables. I think it is evident that the sets of planes which actually

TABLE II. PLANES FOR WHICH L.C.M. OF p AND $r=2$, $l=a$,
$\lambda=2an^2$

p	q	r	s	a/λ	Intensity	h_1	h_2	h_3
1	1	2	4	3	•	2	4	2
1	1	2	8	9	*	2	8	2
1	1	2	12	19	Invisible	2	12	2
2	4	2	0	2·5	Invisible	4	0	2
2	4	2	4	4·5	•	4	4	2
2	4	2	8	10·5	• ?	4	8	2
1	3	2	0	5	*	6	0	2
1	3	2	4	7	*	6	4	2
1	3	2	8	13	• ?	6	8	2
2	8	2	0	8·5	•	8	0	2
2	8	2	4	10·5	• ?	8	4	2
1	5	2	0	13	Invisible	10	0	2

reflect spots can be arranged in a very complete series with few or no gaps. Though at first sight it may appear that in the tables the parameters are selected in a somewhat arbitrary way, they are in reality the simplest possible. For instance, in Table III the first values for p, q, r, s considered are 1, 1, 3, 5. This is so because '$r+s$' must be positive. If $r = 1$, s must be odd, 1, 1, 3, 1 and 1, 1, 3, 3 would reflect the beam so as to miss the photographic plate. 1, 1, 3, 5 and 1, 1, 3, 7 are considered.

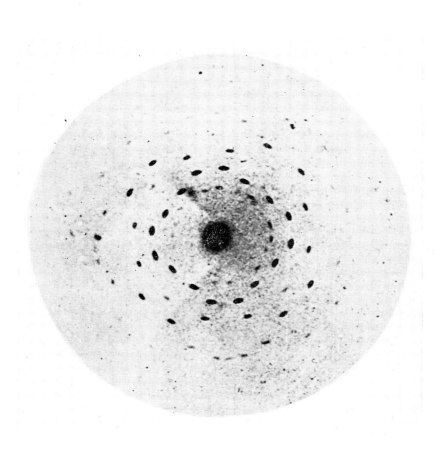

PLATE I.

1, 1, 3, 9 has already been considered as 1, 1, 1, 3, and 1, 1, 3, 11 gives a value for the wavelength outside the 'visible' range.

In Plate I is given a photograph of the interference pattern which Laue obtained. In Fig. 3 the key to the pattern has been drawn, showing in what planes the spots are to be considered as reflected.

Consider a reflecting plane which passes through the atom at the origin and a neighbouring atom, let us suppose the atom whose coordinates are a, o, a. As the plane is turned about the line through these two points the reflected beam traces out a circular cone, which has for axis the line joining the two points and for one of its generators the incident beam. This cone cuts the photographic plate in an ellipse. If the atom through which the plane passes is in the xz plane as above, the ellipse touches the y axis on the photographic plate at the origin. Now take a plane passing through the origin and a point $0, a, 3a$. The locus of the reflected spot as it turns is again an ellipse, which now touches the x axis. The intersections of the two ellipses will give the position of a spot reflected by a plane passing through all three points, the origin, the point $a, 0, a$, and the point $0, a, 3a$.

The ellipses are drawn in the figure, and the plane corresponding to any spot can be found by noting the ellipses at the intersection of which the spot lies. Only those ellipses have been drawn which give the points in Table I. It will be seen that a very large proportion of the spots in the photograph lie at the intersection of these.

The analysis involved in this way of regarding the interference phenomena must be fundamentally the same as that employed by Laue. In Fig. 1, suppose the phase difference between vibrations from successive atoms along the three axes, when waves of wavelength λ fall on the crystal, to be $2\pi h_1, 2\pi h_2, 2\pi h_3$. Then in order that the vibrations from those atoms, which are arranged in the figure at the centres of the cube faces, should also be in phase, one must have

$$h_1/2 - h_3/2 = \text{an integer}, \quad h_2/2 - h_3/2 = \text{an integer}.$$

E

This condition is simply expressed by saying that h_1, h_2, h_3 must all be even or all odd integers. When h_1, h_2, h_3 are given, the value of λ follows from

$$\frac{\lambda}{2a} = \frac{2h_3}{\sqrt{h_1{}^2 + h_2{}^2 + h_3{}^2}},$$

since here $2a$ has been taken as the distance between neighbouring molecules along the three axes.

TABLE III. PLANES FOR WHICH L.C.M. OF p AND $r=3$, $l=2a/3$, $\lambda=4an^2/3$

p	q	r	s	a/λ	Intensity	h_1	h_2	h_3
1	1	3	5	3·6	Invisible	3	5	3
1	1	3	7	5·6	•	3	7	3
1	1	3	11	11·6	Invisible	3	11	3
3	5	3	5	4·9	Invisible	5	5	3
3	5	3	7	6·9	*	5	7	3
3	5	3	11	12·9	Invisible	5	11	3
3	7	3	1	4·9	Invisible	7	1	3
3	7	3	5	6·9	*	7	5	3
3	7	3	7	8·9	*	7	7	3
3	7	3	11	14·9	Invisible	7	11	3
1	3	3	1	7·6	*	9	1	3
1	3	3	5	9·6	+	9	5	3
1	3	3	7	11·6	Invisible	9	7	3

Range of values of a/λ, 5·6—9·6.

TABLE IV. PLANES FOR WHICH THE L.C.M. OF p AND $r=4$, $l=2a/4$, $\lambda=4an^2/4=an^2$

p	q	r	s	a/λ	Intensity	h_1	h_2	h_3
1	1	4	10	8·2	Invisible	4	10	4
1	1	4	14	16·2	Invisible	4	14	4
2	4	4	10	11·2	+	8	10	4
1	3	4	6	12·2	Invisible	12	6	4

If the three simplest values of h_1, h_2, h_3 for a spot on the plate are not all odd, or all even, then these numbers must be doubled to make them even and the wavelength accordingly halved.

When this is done, it can be seen that for each value of h_3 there is a series of values of h_1 and h_2. These numbers all give spots in the photograph if the corresponding value of a/λ lies within a certain range. The smaller the number h_1, the larger is the range of a/λ for which spots are visible. Spots whose a/λ lies near the extremity of the range are very faint, those whose a/λ is in the middle of the range are intense. In the tables the values of h_1, h_2, h_3 corresponding to each spot are set down.

It is quite probable that the qualitative explanation put forward here to account for the intensities of the spots is not the right one, other explanations being possible. For instance, one might substitute for the factor termed 'effective density' above, one which expressed the fact that, other things being equal, spots nearer the centre of the pattern were more intense than those farther out. This, together with the right curve for the distribution of energy in the spectrum of the incident radiation, could be made to account for the intensities quite reasonably. This does not vitiate the conclusion that the spots in the pattern represent a series which is complete, and characteristic of a cubical crystalline arrangement. The other arrangements of cubical point systems cannot, as far as I can see, give such a complete series. The other possible arrangements have for elements of their pattern (1) a cube with a molecule or atom at each corner, the arrangement which Laue pictured, or (2) a cube with a molecule at each corner and one at the centre. Neither arrangement will fit the system of planes given above. It is only the third point system, the element of whose pattern has a molecule at each corner and one at the centre of each cube face, which will lend itself to the system of planes found to represent spots in the photograph.

This last system, seeing that it forms an arrangement of the closest possible packing, is according to the results of Pope and Barlow the most probable one for the cubic form of zinc sulphide.

In one of the photographs taken by Messrs. Friedrich and Knipping the crystal was so oriented that the direction of the incident radiation made equal angles with the three rectangular axes of the crystal. In this case a figure is obtained in which the pattern is a repetition of the spots contained in a sector of angle $\pi/6$. Regarding the spots as reflections of the incident beam in planes as before, these planes can be found almost as easily as those which reflect the spots in the square pattern, and indeed in many cases the planes are identical. I will not give the calculations here, but one point is of especial interest. A photograph was taken of the crystal oriented so that the pattern obtained was perfectly symmetrical. The crystal was then tilted through 3° about a line perpendicular to the incident beam and to one of the cubical axes. This distorted the pattern considerably, but corresponding spots in the two patterns are easily to be recognised. The points which I wish to consider especially are the following.

In the first place, the spots in the distorted pattern are all displaced exactly as would be expected if they were reflections in planes fixed in the crystal. For instance, when the reflecting plane contains the line, about which the crystal was tilted through 3°, it can be ascertained that the movement of the spot corresponds to a deviation of the reflected beam through 6°. This alone is, I think, strong evidence that the wavelength λ is elastic, and not confined to a few definite values, and that equations (1) are satisfied rigorously and not merely approximately.

Besides the distortion of the figure due to the tilting of the crystal, a very marked alteration in the intensity of the spots is to be noticed. This is especially marked for those spots which are near the centre of the pattern, but not on or near the axis about which the crystal is tilted. This is probably due to the fact that for these spots a considerable change in wavelength has taken place.

When the angle of incidence θ of the primary beam on a set of reflecting planes varies, the value of $2d \cos \theta$ is altered and the alteration for the same $\delta\theta$ is greater the greater θ is.

One spot in particular changes from being hardly visible in the symmetrical pattern to being by far the most intense when the crystal is tilted.   It is the spot reflected in a plane passing through the origin and

$$3a, \ 0, \ a; \quad 0, \ 3a, \ a.$$

Planes parallel to this have for d, the shortest distance between successive planes, the value $4a/\sqrt{11}$. It can easily be calculated from the position of the spot that the value of cos θ changes from ·19 to ·12 when the crystal is tilted. This corresponds to a change in the value of a/λ from 4·3 to 6·5, and it was found before for the square pattern that spots corresponding to the former wavelengths were weak, those corresponding to the latter intense.

A curious feature of the photographs may be explained by regarding the spots as formed by reflection. As the distance of the photographic plate from the crystal is altered, the shape of each individual spot varies. At first round, they become more and more elliptical as the plate is moved further away. A reason for this is found in the following. If the incident beam is not perfectly parallel, but slightly conical, rays will strike the crystal at slightly different angles. Regard the crystal as a set of reflecting planes perpendicular to the plane of the paper (Fig. 2). The ray striking the reflecting planes on the upper part of the crystal on the whole meet them at a less angle of incidence than those striking the planes at the bottom; the latter are deflected more, and the rays tend on reflection to come to a focus in a horizontal line. On the other hand, rays deviating from the axial direction in a horizontal plane diverge still more after reflection. Thus as the plate is removed from the crystal, the spots up to a certain distance become more and more elliptical.

The atoms of a crystal may be arranged in 'doubly infinite' series of parallel rows, as well as in 'singly infinite' series of planes. The incident pulse falls on atom after atom in one of these rows, if the row is not parallel to the wave front, and secondary waves are emitted, one from each atom, at definite time intervals. Along any direction lying on a certain circular cone with the row of atoms as axis, these secondary waves will be all in phase, one generator of the cone being, of course, parallel to the direction of the incident radiation. If the row of atoms makes a small angle with the direction, this cone with vertex at the crystal slip may now be considered to cut the photographic plate in an almost circular ellipse passing through the big central spot. Drawing the ellipses which correspond to the most densely packed rows of the crystal, a spot is to be expected at the intersection of two ellipses, for this means that pulses from a doubly

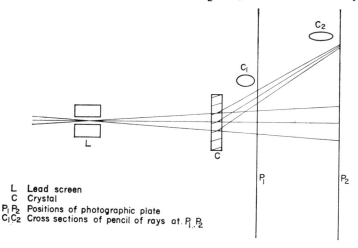

L   Lead screen
C   Crystal
$P_1 P_2$ Positions of photographic plate
$C_1 C_2$ Cross sections of pencil of rays at $P_1 P_2$

Fig. 2.

infinite set of atoms are in that direction in agreement of phase. Thus it ought to be possible to arrange the spots in the photograph on these ellipses, in whatever way the crystal is oriented, and indeed they appear in all cases. They come out very strongly in the photographs taken with copper sulphate crystals.

So far it has been assumed that the atoms of zinc and sulphur act in an identical manner with regard to the production of secondary waves, but this assumption is not necessary. What is brought out so strongly by the analysis is this; that the point system to be considered has for element of its pattern a point at each corner of the cube and one at the centre of each cube face. In the arrangement assigned to cubical zinc sulphide and similar crystals by Pope and Barlow, this point system is characteristic of both the arrangement of the individual atoms regarded as equal spheres, and of the arrangement of atoms which are in every way identical as regards nature, orientation, and neighbours in the pattern. The atoms of zinc, for instance, in the zinc blende are grouped four together tetrahedron-wise, and as these little tetrahedra are

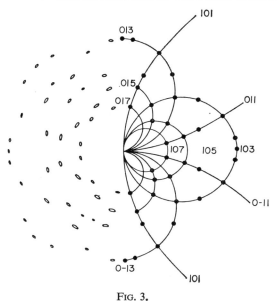

FIG. 3.

all similarly oriented and are arranged themselves in the above point system, atoms of zinc identical in all respects will again be arranged in this point system. Which of these factors it is that decides the form of the interference pattern might be found by experiments with crystals in which the point system formed by the centres of all the atoms differs from that formed by the centres of identical atoms.

In conclusion, I wish to thank Professor Pope for his kind help and advice on the subject of crystal structure.

3. The Crystalline Structure of Copper*

W. Lawrence Bragg†

The copper crystals used in this investigation were some natural specimens, for which I am indebted to Mr. Hutchinson, of the Mineralogical Laboratory at Cambridge. In their natural state these specimens are obviously rough crystals, and have some faces of large dimensions (1 cm. each way), but these faces are very much warped and distorted. An attempt was made to obtain an X-ray reflexion from various natural faces of such crystals, but it was not successful. Apparently the outer surface of the crystal has been so battered and distorted that little regular crystalline arrangement is left. Any attempt to grind crystal faces artificially also destroys the crystalline character of the surface and so prevents the reflexion of X-rays from the face.

It was observed, however, that when the crystal was placed in nitric acid until the surface was eaten away to an extent of perhaps $\frac{1}{4}$ millimetre, the faces were etched deeply into numerous parallel facets, which all reflected the light simultaneously in the usual way. This suggested that, internally, the crystal structure was perfect, and showed further that in some cases the whole specimen was composed of a single crystal. Moreover, in this case the surface layers are not pulled about, and so are capable of reflecting the X-rays falling on them. This method of obtaining a crystal surface was suggested by some previous experiments on

*_Philosophical Magazine_, Vol. **28** (2), pp. 355–60 (1914).

†Allen Scholar of the University of Cambridge. Communicated by the Author.

natural zinc oxide, zincite. Zincite occurs very rarely as crystals, and the specimens used had merely a platy structure. However, by partly dissolving a block of the mineral in hydrochloric acid, the etched mass showed indications of crystalline structure sufficient to serve as a guide in the preparation of various faces. The faces reflected the X-rays and led to the determination of the arrangement of zinc and oxygen atoms.

Copper crystallizes in the cubic system, holohedral class. The natural crystals of copper used in the experiments were mostly of one type, being composed of two individuals twinned about the plane (111). The faces of the simplest crystals approximated to those of the rhombic dodecahedron $\{110\}$. The face first investigated was that parallel to the twin plane (111) of the best of the crystals. The crystal appeared like two triangular pyramids joined base to base, and the apex of one pyramid was ground down on a carborundum wheel until a triangular face (111) was formed. This was roughly polished, treated with nitric acid to dissolve away the outer layers, and then mounted in the usual way in the X-ray spectrometer.[1]

By assuming various arrangements for the copper atoms, we can calculate the spacings of various planes of the crystal, starting from the density of copper, 8·96, as a basis. For instance, if the copper atoms are arranged at the corners of cubes, so as to form a simple cubic lattice, we have the relation

$$(d_{(100)})^3 . 8 \cdot 96 = 63 \cdot 57 . 1 \cdot 64 . 10^{-24},$$

since we know that

1. Mass of a copper atom = 63·57.(Mass of a hydrogen atom)

$$= 63 \cdot 57 . 1 \cdot 64 . 10^{-24} \text{ gram},$$

2. The unit cube of the structure contains one copper atom. This gives the relation

$$d_{(100)} = 2 \cdot 26 . 10^{-8} \text{ cm}.$$

---

[1] For a description of the instrument and its manipulation, see *Proc. Roy. Soc.* A, **88**, 428, and A, **89**, 468.

The X-rays from an anticathode of palladium, such as was used for the purposes of this experiment, have a wavelength of $\cdot576 . 10^{-8}$ cm. This is reflected from the planes $d_{(100)}$ at an angle given by the equation

$$\lambda = 2 . d_{(100)} . \sin \theta$$

Substituting the above values for λ and $d_{(100)}$, we find that $\theta = 7° 20'$.

Suppose, on the other hand, that the copper atoms were on a face-centred cubic lattice. This lattice has a point at each corner of a set of cubes, and one at the centre of each cube face. The volume $(d_{(100)})^3$ now contains only one-half a copper atom. From this it can be calculated that $d_{(100)} = 1 \cdot 80 . 10^{-8}$ cm. So for the other faces; the results are summarized below.

	Spacing of planes.	Glancing angle of reflexion, *Pd* rays.
Face-centred lattice:		
	$d_{(100)} = 1 \cdot 80 . 10^{-8}$ cm.	$\theta_{(100)} = 9° 13'$
	$d_{(110)} = 1 \cdot 27 . 10^{-8}$ cm.	$\theta_{(110)} = 13° 2'$
	$d_{(111)} = 2 \cdot 08 . 10^{-8}$ cm.	$\theta_{(111)} = 8° 0'$
Simple cubic lattice:		
	$d_{(100)} = 2 \cdot 26 . 10^{-8}$ cm.	$\theta_{(100)} = 7° 20'$
	$d_{(110)} = 1 \cdot 59 . 10^{-8}$ cm.	$\theta_{(110)} = 10° 22'$
	$d_{(111)} = 1 \cdot 30 . 10^{-8}$ cm.	$\theta_{(111)} = 12° 50'$

The face (111) of the copper crystal, prepared as described above, was first investigated. The chamber was set at $25° 40'$ ($= 2 \times 12° 50'$). The crystal face, being adjusted so that the rays fell on it at a small glancing angle, was turned so that this angle assumed in turn a series of values between 6° and 20°, in order to find if, in some position, it reflected the X-rays. This had to be done because the true orientation of the plane (111) in the crystal was not known with any exactness. At none of these angles was there a reflexion into the chamber, the simple cubic arrangement of the copper atoms being therefore ruled out.

On setting the ionization chamber at 16°, however, a marked effect was found. Fig. 1 shows the current in the chamber for a series of angles at which the crystal was set. Between 8° and 11° 30′ there is a marked increase in the ionization current, which rises to a maximum at 9° 30′. Now if the crystal were perfect, the range of angles at which it reflected the X-rays would be limited to some 30′ at most. The fact that the crystal reflects

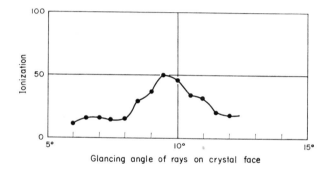

FIG. 1.

over such a wide range of angles shows that its planes are distorted to an extent of several degrees, instead of being strictly parallel. As the crystal is turned round, one set of planes after another comes into the reflecting position and causes an ionization current in the chamber. From the curve we deduce that when the crystal is set at 9° 30′ the area of its face so oriented as to reflect is larger than at any other angle.

A series of readings are now taken with the chamber at various angles "θ", and the crystal in each case at the angle "$\theta/2 + 1° 30′$", in order always to make use of this larger reflecting area. The results are shown in the curve for the face (111) of Fig. 2. Here we have two peaks close together in the curve, representing the two lines in the spectrum of palladium. The curve shows a decided first- and second-order spectrum, and even perhaps a third, though this last is somewhat doubtful. It must be noted that the range of angles, over which the ionization chamber may be set so as to receive the reflected beam, is confined to 1°, though the crystal structure is so imperfect. If the crystal were perfect it would be scarcely smaller, and the reason for this is clear. Although there are a number of settings for the crystal which

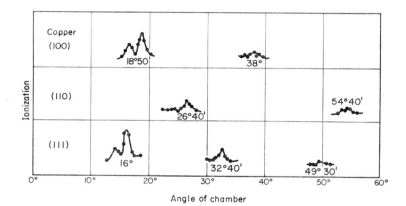

Angle of chamber

Fig. 2.

enable a set of planes somewhere on its distorted face to receive the incident X-rays at a glancing angle of 8° and so to reflect them, the reflected rays all converge and are received by the chamber when set exactly at 16°.[2] Therefore, when the chamber and crystal are moved simultaneously, the reflexion is only found when the chamber is in the neighbourhood of 16°.

Figure 2 also shows the curves for faces (110) and (100) of copper. The angles at which the spectra are found are as follows:—

$\theta_{(100)} = \ 9° \ 24' \sin \theta_{(100)} = \cdot163.$ Calculated angle 9° 13'
$\theta_{(110)} = 13° \ 18' \sin \theta_{(110)} = \cdot230$,,   ,, 13° 2'
$\theta_{(111)} = \ 8° \ 0' \ \sin \theta_{(111)} = \cdot139.$,,   ,, 8° 0'
$\sin \theta_{(100)}: \ \sin \theta_{(110)}: \ \sin \theta_{(111)} = 1:1\cdot41:\cdot85 = 1:\sqrt{2}:\sqrt{3}/2$

[2]Cf. W. H. Bragg, *Phil. Mag.* May 1914, p. 887.

This relation between the angles of reflexion from the three principal faces of the crystal is that which would exist for a face-centred cubic lattice. In such a lattice

$$d_{(100)}:d_{(110)}:d_{(111)} = 1:1/\sqrt{2}:2/\sqrt{3}.$$

We have already seen that the absolute values of the angles at which reflexion occurs are those to be expected if the copper atoms lie on a face-centred lattice. It is further to be observed that the 1st, 2nd, and 3rd order peaks reflected from the faces of the crystals are in every case quite normal, the first order being greater than the second, and that greater than the third. This, as has been shown in former papers (*Proc. Roy. Soc.* A. vol. lxxxix. p. 472), implies a regular arrangement of reflecting planes, equally spaced and identical in nature. Lastly, as a check, a search was made for spectra at half the angles at which the first spectra of Fig. 2 occur, in the case of the planes (100) and (110). This search gave a negative result. Taking all this into consideration, there can be little doubt that *the atoms of a copper crystal are arranged on a face-centred cubic lattice.*

The results are interesting in that they show that considerable accuracy of measurement can be obtained with the X-ray spectrometer, even when the crystal itself is highly irregular. If the cubic symmetry of the copper crystal had not rendered this unnecessary, it would have been possible to measure the axial ratio of the crystal within 1 per cent. of the truth, although the faces of the crystal were distorted by many degrees. Since the crystal is so irregular, only a fraction of its surface reflects at any one angle, and therefore the electroscope had to be very sensitive when measuring the small ionization current. This explains the very obvious irregularities of the curves in Fig. 2. The dots in this figure represent a set of readings. All these curves were repeated several times, some with different crystal faces. Some curves were more irregular than others, but all agreed closely in the positions of the spectra.

I wish to take this opportunity of again thanking Mr. Hutchinson for his kind help, both in supplying material and in aiding with his advice the preparation of the various crystal faces. I wish to thank Professor Sir J. J. Thomson for his kind interest in the experiments. I am indebted to the Institut International de Physique Solvay for a grant with the aid of which the apparatus used in these experiments was purchased.

Summary

It was found possible, by treating with acid prepared surfaces of a natural crystal of copper, to obtain crystal faces which could be used as reflectors in the X-ray spectrometer.

The results of the investigations, thus rendered feasible, showed that in a copper crystal the atoms are arranged on a face-centred cubic lattice. This is the close-packed lattice, to which attention has been drawn by Pope and Barlow. The crystal structure is the most simple of any as yet analysed.

The Cavendish Laboratory, July 16th, 1914.

4.   A New Method of X-ray Crystal Analysis*

A. W. HULL

THE beautiful methods of crystal analysis that have been developed by Laue and the Braggs are applicable only to individual crystals of appreciable size, reasonably free from twinning and distortion, and sufficiently developed to allow the determination of the direction of their axes. For the majority of substances, especially the elementary ones, such crystals cannot be found in nature or in ordinary technical products, and their growth is difficult and time-consuming.

The method described below is a modification of the Bragg method, and is applicable to all crystalline substances.   The quantity of material required is preferably ·005 c.c., but one tenth of this amount is sufficient. Extreme purity of material is not required, and a large admixture of (uncombined) foreign material, twenty or even fifty per cent., is allowable provided it is amorhpous or of known crystalline structure.

Outline of Method

The method consists in sending a narrow beam of mono-chromatic X-rays (Fig. 2) through a disordered mass of small crystals of the substance to be investigated, and photographing the diffraction pattern produced. Disorder, as regards orientatiog

---

Physical Review, **10**, 661 (1917). A brief description of this method was given before the American Physical Society in October, 1916, and published in the *Physical Review* for January, 1917.

of the small crystals, is essential. It is attained by reducing the substance to as finely divided form as practicable, placing it in a thin-walled tube of glass or other amorphous material, and keeping it in continuous rotation during the exposure.[1] If the particles are too large, or are needle-shaped or lamellar, so that

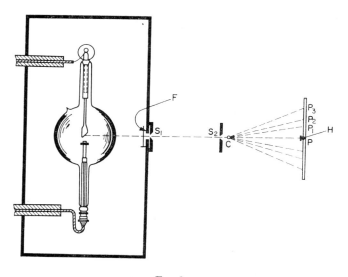

Fig. 2.

they tend to assume a definite orientation, they are frequently stirred. In this way it is assured that the average orientation of the little crystals during the long exposure is a random one. At any given instant there will be a certain number of crystals whose 100 planes make the proper angle with the X-ray beam to reflect the particular wavelength used, a certain number of others whose 111 planes are at the angle appropriate for reflection by these planes, and so for every possible plane that belongs to the

---

[1] If the powder is fine, rotation is not necessary unless great precision is desired. With crystal grains ·01 cm in diameter, or less, the pattern generally appears quite uniform without rotation.

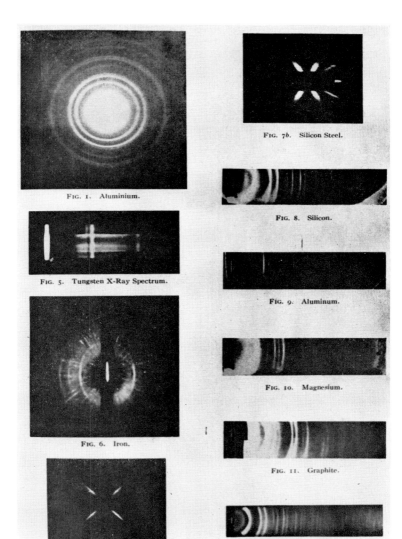

FIG. 1. Aluminium.

FIG. 5. Tungsten X-Ray Spectrum.

FIG. 6. Iron.

FIG. 7a. Silicon Steel.

FIG. 7b. Silicon Steel.

FIG. 8. Silicon.

FIG. 9. Aluminum.

FIG. 10. Magnesium.

FIG. 11. Graphite.

FIG. 12. Diamond.

PLATE I.

crystal system represented. Each of these little groups will contain the same number of little crystals, provided the distribution is truly random, and the total number of crystals sufficiently large. This condition is very nearly realized in the case of fine powders, and may, by sufficient rotation and stirring, always be realized for the *average orientation* during the whole exposure; that is, there will be, *on the average*, as many cubic centimeters of crystals reflecting from their 100 planes as there are cubic centimeters reflecting from 111, 210, or any other plane. This is true for every possible plane in the crystal.

The diffraction pattern should contain, therefore, reflections from every possible plane in the crystal, or as many of these as fall within the limits of the photographic plate. Figure 1, Plate I, shows the pattern given by aluminium when illuminated by a small circular beam of nearly monochromatic rays from a molybdenum tube. The exposure was nine hours, with 37 milliamperes at 30,000 volts, and crystal powder 15 cm. from the target and 5·9 cm. from photographic plate. The faintness of the vertical portions of the circles is due to the cylindrical form in which the powder was mounted, causing greater absorption of rays scattered in the vertical plane. Patterns containing many more lines are shown in Figs. 6–10, where the diaphragm limiting the beam was a slit instead of a circular aperture, and the pattern was received on a photographic film bent in the arc of a circle.

The number of possible planes in any crystal system is infinite. Hence if equal reflecting opportunity meant equal reflected energy, it would follow that the energy reflected by each system of planes must be an infinitesimal fraction of the primary beam, and hence could produce no individual photographic effect. It is easily seen, however, that only those planes whose distance apart is greater than $\lambda/2$, where λ is the wavelength of the incident rays, can reflect any energy at all. Planes whose distance apart is less than this cannot have, in any direction, except that of the incident beam, equality of phase of the wavelets diffracted by electrons in consecutive planes. Hence the resultant amplitude associated with any such plane is very small, and would be identically zero for a perfect lattice and sufficiently large number of planes. The total scattered energy is therefore divided among a finite number of planes, each of which produces upon the photographic plate a linear image of the source (cf. Fig. 1). The total possible number of these lines depends upon the crystal structure and the wavelength. For diamond, with the wavelength of the K_a doublet of molybdenum, $\lambda = 0.712$, the total number of lines is 27. All of these are present in the photograph shown in Fig. 12. For the rhodium doublet, $\lambda = 0.617$, the total number is 30; for the tungsten doublet, $\lambda = 0.212$, it is more than 100; while the iron doublet, $\lambda = 1.93$, can be reflected by only three sets of diamond planes, the octahedral (111), rhombic dodecahedral (110), and the trapezohedral (311). The diffraction pattern in this case would consist, therefore, of but three lines.

The positions of these lines, in terms of their angular deviation from the central beam, are completely determined by the spacing of the corresponding planes, according to the classic equation $n\lambda = 2d \sin \theta$, where θ is the angle between the incident ray and the plane, hence 2θ is the angular deviation, d the distance between consecutive planes, λ the wavelength of the incident rays, and n the order of the reflection. The calculation of these positions is discussed in detail below.

The relative intensity of the lines, when corrected for temperature, angle, and the number of cooperating planes, depends only upon the space distribution of the electrons of which the atoms are composed. Most of these electrons are so strongly bound to their atoms that their positions can probably be completely specified by the positions of the atomic nuclei and the characteristic *structure* of the atom. Experiments are in progress to determine such a structure for some of the simpler atoms. A few of the electrons, however, are so influenced by the proximity of other atoms, that their position will depend much on the crystal structure and state of combination of the substance. There is also good reason to believe that certain electrons are really *free*, in that they belong to no atom, but occupy definite spaces in the lattice, as though they were atoms.

With elements of high atomic weight, where each atom contains a large number of electrons, the majority of these electrons must be quite close to the nucleus, so that the intensity of the lines will depend primarily upon the position of the nuclei relative to their planes, and only slightly upon the characteristic structure of the atom and the position of valence and free electrons. With these substances, therefore, the relative intensity of the lines gives direct evidence regarding the positions of the atoms, and may be used, in the manner described by the Braggs,[2] for the determination of crystal structure. The powder photographs have an advantage, in this respect, over ionization-chamber measurements, in that the intensities of reflection from different planes, as well as different orders, are directly comparable, which is not true of ionization-chamber measurements unless the crystal is very large and may be ground for each plane.

In the case of light substances, on the other hand, the intensities depend very much on the internal structure of the atoms, and unless this structure is known or postulated, but little weight should be given to intensity in determining the crystal structure. Much evidence for the structure of these elements may be obtained, however, from the observation of the *position* of a large number of lines, and this evidence will generally be found sufficient. The examples given at the end of this paper are all elements of low atomic weight, and the analysis given is based entirely on the position of the lines. The photographs used for the analysis are preliminary ones, taken with very crude experimental arrangements, and yet in every case except one the evidence is sufficient.

The method of measuring and interpreting intensity will form the subject of a future paper.

Experimental Arrangement

The arrangement of apparatus is shown in Fig. 2. The X-ray tube is completely enclosed in a very tightly built lead box. If a tungsten target is to be used this box should be of $\frac{1}{4}$ inch lead,

---

[2] *X-Rays and Crystal Structure*, pp. 120 ff.

with an extra $\frac{1}{4}$ inch on the side facing the photographic plate. If a rhodium or molybdenum target is used $\frac{1}{8}$ inch on the side toward the photographic plate, and $\frac{1}{16}$ inch for the rest of the box, is sufficient. The rays pass through the filter F and slits S_1 and S_2, and fall upon the crystal substance C, by which they are diffracted to points p_1, p_2, etc., on the photographic plate P. The direct beam is stopped by a narrow lead strip H, of such thickness that the photographic image produced by this beam is within the range of normal exposure. For a tungsten target, the thickness of this strip should be $\frac{1}{8}$ inch; for a molybdenum target about 1/100 inch.

The X-ray Tube

In order to produce monochromatic rays, it is necessary to use a target which gives a characteristic radiation of the desired wavelength, and to run the tube at such a voltage that the radiation of this wavelength will be both intense and capable of isolation by filtering.

The relation between general and characteristic radiation at different voltages has been investigated, for tungsten and molybdenum, by the author,[3] and, in more detail, for rhodium by Webster and platinum by Webster and Clark.[4] The results may be summarized as follows: The characteristic *line spectra* are excited only when the voltage across the tube is equal to or greater than the value $V = hv/e$, where h is Planck's constant, e the charge of an electron, and v the frequency corresponding to the short wavelength limit of the series to which the line belongs.

With increase of voltage above this limiting voltage, the intensity of the lines increases rapidly, approximately proportional to the 3/2 power of the excess of voltage above the limiting value.[5] The following table will show the rate of increase for

[3] *Nat. Acad. Proc.* **2**, 268 (1916).

[4] *Phys. Rev.* **7**, 599, 1916; *Nat. Acad. Proc.* **3**, 185 (1917).

[5] Webster and Clark, *Proc. Nat. Acad.* **3**, 185 (1917).

the α line of the K series of molybdenum, as used in the experiments described below.[6]

TABLE I.   INCREASE OF INTENSITY OF THE K_a LINE OF Mo WITH VOLTAGE

	Kilovolts									
	20	22	24	26	28	30	32	34	36	40
Intensity	0	1·25	2·75	4·80	7·30	9·60	12·65	15·2	18·5	23·4

The rapid increase of characteristic radiation with voltage makes it desirable to use as high voltage as possible. If the voltage is too high, however, a part of the general radiation, whose maximum frequency is directly proportional to the voltage,[7] becomes so short that it is impossible to separate it from the characteristic by a selective filter. With a molybdenum target the best working voltage is about 30,000 volts, with tungsten about 100,000 volts.

Filters

Although it is impossible to produce truly monochromatic radiation by filtering, it is easy to obtain a spectrum containing only *one line*, and in which the intensity of this line is more than thirty times that of any part of the general radiation.   To accomplish this, use is made of the sudden increase in absorption of the filter at the wavelength corresponding to the limit of one of its characteristic series; that is, at the wavelength which is just short enough to excite in the filter one of its characteristic radiations.   A filter is chosen whose K series limit[8] lies as close

---

[6]The general radiation of the same wavelength as the α line is included in these values.

[7]Duane and Hunt, see *Phys. Rev.* 6, 619, and Hull, *ibid.* 7, 156.

[8]A complete table of wavelengths of series lines for all elements thus far investigated is given by Siegbahn, *Ber. d. D. Phys. Gesel.* 13, 300 (1917).

as possible to the desired wavelength *on its short wavelength side*. For example, to isolate the K lines of molybdenum whose wavelength is ·712 Å., the most appropriate filter is zirconium, the limit of whose K series is at $\lambda = ·690$ Å. The absorption coefficient of the filter is then a minimum for the wavelength in question, and increases rapidly with wavelength *in both directions*; on the left, toward shorter wavelengths, it jumps suddenly by

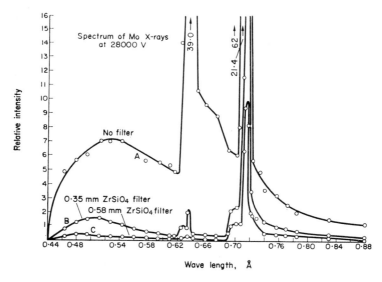

FIG. 3.

about 8-fold; on the right it increases more slowly, viz., as the cube of the wavelength.[9]

If the longest wavelength in the series, which, fortunately, in the case of the K series, is the most intense, is chosen for the monochromatic ray, the eightfold increase in absorption coefficient will completely eliminate the other lines of the series, while

[9]Hull and Rice, *Phys. Rev.* **8**, 326 (1916).

reducing the chosen line by only one-half. To eliminate the general radiation is not so easy. Webster has shown[10] that the intensity of the characteristic radiation increases more rapidly with voltage than that of the neighbouring general radiation, so that the higher the voltage the more prominently the line stands

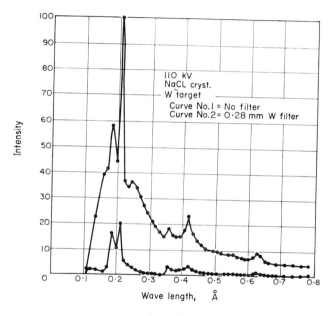

110 kV
NaCL cryst.
W target
Curve No.1 = No filter
Curve No.2 = 0·28 mm W filter

Fig. 4.

out above adjacent wavelengths, and this is the only way in which it can be sharply separated *from longer wavelengths.* If the voltage is too high, however, the shortest wavelength end of the general spectrum becomes transmissible by the filter, and while its wavelength is far removed from that of the line which is to be isolated, and it can itself produce no line image, yet its integral effect produces a general blackening of the plate that obscures

[10]L.c.

the lines. Sharp limitation on the *short* wavelength side is obtained by the selective action of the filter.

It is necessary, therefore, to choose filter material, filter thickness, and voltage, to correspond to the target used. For a molybdenum target, the filter should be zirconium, and a thickness of about 0·35 mm. of powdered zircon is sufficient[11] (see Fig. 3). The optimum voltage is between 28,000 and 30,000 volts. For a tungsten target the filter should be ytterbium, of a thickness of about 0·15 mm., but this has not yet been tested. A filter of this thickness of metallic tungsten or tantalum eliminates most of the general spectrum, but leaves the β doublet as well as the α doublet, which is very undesirable (cf. Figs. 4 and 5). The optimum voltage is about 100,000 volts.

The effect of filtering on the spectrum of a molybdenum target at 28,000 volts is shown in Fig. 3, which gives the intensity of the different wavelengths as measured with an ionization chamber, so constructed as to eliminate, nearly, errors due to incomplete absorption.[12] No correction has been made for coefficient of reflection of the (rock salt) crystal. The intensities of the K lines are too great to be shown on the figure, the α line being four times and the β line two and one half times the height of the diagram. A filter of ·35 mm. of zircon reduces the intensity of the α line from 62 to 21·4; while reducing the β line from 39 to 2·2. The general radiation to the left is still quite prominent. An increase in filter thickness from ·35 mm. to ·58 mm. (Curve C) reduces it but little more than it reduces the α line, so that very

---

[11] The absorption of the Si and O in zircon is negligible compared to that of the zirconium, so that crystal zircon is as efficient as metallic zirconium.

[12] The ionization chamber contains two electrodes of equal length. The second electrode, the one farther from the crystal, was connected to the electrometer, and the pressure of methyl odide in the chamber was such that the wavelengths in the middle of the range investigated suffered 50 per cent. absorption in passing through the first half of the chamber. The electrometer deflection is proportional to $I_0 e^{-\mu l}(1 - e^{-\mu l})$, where I_0 is the intensity on entering the chamber, l the length of either electrode and μ the coefficient of absorption of the methyl odide. This expression has a very flat maximum for $e^{-\mu l} = \frac{1}{2}$, so that for a considerable range on either side, the readings are proportional to I_0.

little is gained by additional filtering. The sudden increase in absorption of the zirconium is seen at $\lambda_0 = 0\cdot690$ Å., which is exactly the short wavelength limit of its K series, as extrapolated from Malmer's values of the β_1 and β_2 lines of yttrium and the β_1 line of zirconium.

The effect of a tungsten filter upon the spectrum of tungsten at 110,000 volts is shown in Figs. 4 and 5. Here the critical wavelength of the filter is at the short wavelength edge of the whole series, so that all the lines are present. A filter of ytterbium would eliminate all but the α doublet. Figure 4 gives the ionization chamber measurements, uncorrected, of the tungsten spectrum at 110,000 volts, as reflected by a rock salt crystal. The upper curve is the unfiltered spectrum, the lower that which has passed through a filter of 0·15 mm. of metallic tungsten. The K lines are much more prominent in the filtered than in the unfiltered spectrum, but the general radiation, especially the short wavelength end, is much too prominent, showing that the voltage is too high. In Fig. 5 the effect of the tungsten filter (above) is compared with that of 1 cm. of aluminium (below), in order to show more clearly the selective effect of the tungsten filter. The wide middle portion of the spectrum is unfiltered.

The Crystalline Material

The Bragg method of X-ray crystal analysis is by far the simplest whenever single crystals of sufficient perfection are available. If, however, perfect order of crystalline arrangement cannot be had, the next simplest condition is perfect chaos, that is, a random grouping of small crystals, such that there is equi-partition of reflecting opportunity among all the crystal planes. This has two disadvantages, viz., that the opportunity of any one plane to reflect is very small, so that long exposures are necessary; and the images from all planes appear on the same plate, so that it is impossible, without calculation, to tell which image belongs to which plane. It has the advantages, on the other hand, of allowing a definite numerical calculation of the position and intensity of each line, and of being free from uncertainties due to imperfection and twinning of crystals. In the latter respect it serves as a valuable check on the direct Bragg method.

The crystalline material is, wherever possible, procured in the form of a fine powder of ·01 cm. diameter or less. This may be accomplished by filing, crushing, or by chemical or electrochemical precipitation, or by distillation. In the case of the metals like alkalies, to which none of these methods can be applied, satisfactory results have been obtained by squirting the metal through a die in the form of a very fine wire, which is packed, with random folding, into a small glass tube, and kept in continuous rotation, with frequent vertical displacements, during exposure.

The method of mounting the crystalline substance depends on the wavelength used. If tungsten rays ($\lambda = 0.212$) are used, so that the angles of reflection, for all visible lines, are small, it is most convenient to press the powder into a flat sheet, or between plane glass plates, and place this sheet at right angles to the beam. In this case the correction for the difference in absorption of the different diffracted rays is negligible. If a molybdenum tube is used, on the other hand, diffracted rays can be observed at angles up to 180° (cf. Fig. 10), so that the substance must be mounted in a cylindrical tube. In this case also, the correction for absorption is unnecessary, provided the diameter of the tube is properly chosen and the beam of rays is wide enough to illuminate the whole tube.

The optimum thickness of crystalline material, for a given wavelength, may be calculated approximately as follows:

Let k represent the scattering coefficient and μ the absorption coefficient of the substance for the wavelength used, and I_0 the intensity of the incident rays. The intensity scattered by a thin layer dx at a distance x below the surface will be

$$dR = kI_0e^{-\mu x}dx.$$

This radiation will suffer further absorption in passing through a thickness $t - x$, approximately, where t is the thickness of the sheet. Hence the total intensity of the scattered radiation that emerges will be

$$R = \int_0^t kI_0 e^{-\mu t} dx$$

$$= kI_0 t e^{-\mu t}.$$

This will be a maximum when

$$\frac{dR}{dt} = kI_0(e^{-\mu t} - \mu t e^{-\mu t}) = 0$$

or

$$t = I/\mu,$$

where t is the thickness of the crystalline sheet in centimeters and μ the linear absorption coefficient.

If the material is in cylindrical form, the optimum diameter is slightly greater than the above value.

Exposure

Very long exposures, as remarked above, are necessary if a large number of lines is desired, and it is important to increase the speed by the use of an intensifying screen, and by bringing the crystal as close as practicable to the tube. With rays as absorbable as those from a molybdenum tube, it is necessary to use films, not plates, with the intensifying screen. Under reasonable conditions, an exposure of ten to twenty hours will produce a general blackening of the plate well within the limit of normal exposure. Since a greater density than this cannot increase the contrast, nothing is to be gained by longer exposure. Further detail can be hoped for only by using more nearly monochromatic rays, screening the plate more perfectly from stray and secondary rays in the room, and decreasing the ratio of amorphous to crystalline material in the specimen under examination.

5. On the Interpretation of X-ray, Single Crystal, Rotation Photographs*

J. D. BERNAL

IN THE development of the study of crystals by X-rays the methods used divide themselves naturally into four types: the Bragg Ionisation Spectrometer method, the Laue method, the Powder method of Debye and Scherrer, and the Rotating Crystal method of Rinne, Schiebold and Polyani. The techniques of the first three of these methods are fully explained in such books as 'X-rays and Crystal Structure', by W. H. and W. L. Bragg, 'The Structure of Crystals', by Wyckoff, and 'Krystalle und Röntgenstrahlen', by Ewald, as well as in original papers. On the other hand, the rotation method is only slightly touched on in these works, the literature is scattered in a great number of papers, and the technique has not so far been described at any length in a convenient form. Particularly in English, references to it are scanty.

In this paper the author has tried to give a concise account of the method, together with various types of charts and tables, as it is used in the Davy Faraday Laboratory. The methods described differ in certain respects from those used on the Continent,[1] but they have been found to be rapid and sufficiently accurate.

In the rotation method proper a small crystal, mounted on a spindle which can be revolved at a uniform speed, is rotated in front of a narrow beam of homogeneous X-rays and the beams of

---

*_Proceedings of the Royal Society_, A, **113**, 117 (1927); the first part of the paper, pp. 117–23, is reproduced. Communicated by Sir William Bragg, F.R.S. Received July 24, 1926.
[1]For a convenient summary, see Schiebold, _Z. f. Physik_. **28**, 355 (1924).

X-rays reflected record themselves as spots on a plate or film placed behind the crystal. In the more developed form of the oscillation method, instead of a complete rotation, the spindle is turned backwards and forwards at a constant speed through some definite small angle. It is not within the scope of the present paper to describe the forms of apparatus which may be used for this purpose; they will be touched on only in so far as they have a bearing on the mathematical interpretation of the experimental results.

In the analysis of crystal structures by any of the methods so far used, the experimental data are all comprised in the knowledge of the intensities of the X-ray reflections from every one of the triply infinite series of crystallographic planes in a crystal. Naturally in practice it is only a question of measuring as many of these reflections as possible, especially those of small crystallographic indices. The procedure can be conveniently divided into three stages:—

   (i) the determination of the size and shape of the unit cell of the fundamental lattice of the crystal;

  (ii) the determination of the indices of the reflecting planes leading to the determination of the space group;

 (iii) the measurement of the intensity of the reflection from each plane leading to the complete determination of the structure.

The rotation method is best suited to the first two of these stages, and though its use is being extended to the last, only the first two will be dealt with here.

The understanding of the phenomena of crystal diffraction of X-rays is much helped by a consideration of the reciprocal lattice. This conception was first used by Ewald,[2] to whose paper reference should be made, but the account given here refers more particularly to the rotation method and to graphical applications.

---

[2] *Z. f. Kryst.* **56**, 129 (1921).

We may consider the crystal in the first place as a simple space lattice. Choosing one point of the lattice as origin, we may take the three vectors a, b, c, as three primitive translations which form a primitive triplet. Any plane of this lattice may be represented by the indices h, k, l, where the intercepts of such a plane on a, b, c, are a/h, b/k, c/l, respectively. We will, however, restrict the term "plane" to those planes for which h, k, l, are integers (not

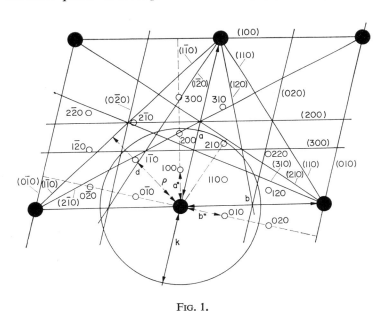

FIG. 1.

● Point of Lattice (310) Plane of Lattice ○310 Point of Reciprocal Lattice.

necessarily prime to each other). In other words, in any set of parallel crystallographic planes, we choose the one nearest the origin and the planes parallel to this whose distances from the origin are sub-multiples of its distance from the origin. The former planes are responsible for the first-order X-ray reflections, the latter for the higher orders.

Now, from the planes of the lattice, as so defined, another lattice can be built up each point of which lies on the normal from the origin to a plane and at a distance ρ from the origin, ρ and the spacing d of the plane being related by

$$\rho d = k^2$$

where k is a constant. In other words, each point in the new lattice is the reciprocal polar of a plane in the old lattice in a sphere of radius k. The new lattice is consequently called the reciprocal lattice of the old. By applying to the new lattice the same operation, we arrive at a replica of the old lattice. For every plane in one there corresponds a point in the other and vice versa (see Fig. 1).

We may refer this reciprocal lattice to a set of vectors a^*, b^*, c^*, reciprocal to the set a, b, c, i.e., such that in general

$$a^* = \frac{k^2 bc}{\Delta} \sin \alpha \text{ and is perpendicular to the plane } bc,$$

$$b^* = \frac{k^2 ca}{\Delta} \sin \beta \text{ and is perpendicular to the plane } ca,$$

$$c^* = \frac{k^2 ab}{\Delta} \sin \gamma \text{ and is perpendicular to the plane } ab,$$

where Δ is the volume of the parallelepipedon abc. A lattice built upon a^*, b^*, c^*, as primitive translations will be the reciprocal lattice of that built up on a, b, c, and vice versa (see Fig. 2). If (h, k, l) are the indices of any plane in the original lattice, then the corresponding point in the reciprocal lattice has for co-ordinates h, k, l.

It is difficult to obtain a physical picture of the phenomena in the reciprocal lattice corresponding to the diffraction of X-rays in crystals. By assuming Bragg's law, however, we can obtain a geometrical one. In the original lattice, if θ is the glancing angle, i.e., the angle between the reflecting plane and the normal to the wave front of the incident beam, reflection only takes place if

$$\sin \theta = \lambda/2d.$$

In the reciprocal lattice the wave fronts reciprocate into points travelling along their normal and θ becomes the angle between the incident ray and the plane through the corresponding point in the reciprocal lattice which is normal to its radius vector. As this radius vector $\rho = k^2/d$ we have as the expression analogous to Bragg's law

$$\sin \theta = \lambda\rho/2k^2.$$

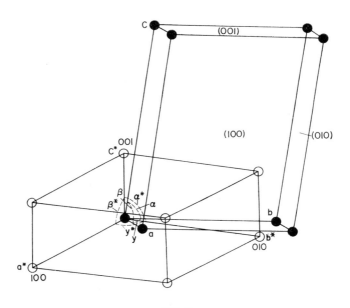

Fig. 2.

The value of the use of the reciprocal lattice is that as every point in it corresponds to a reflection of a plane in the original crystal, it also corresponds to a spot on the photograph, and the arrangement of spots on the photograph is similar to the arrangement of points in the reciprocal lattice.

The reflection of X-rays is shown geometrically in Fig. 3. If we consider a sphere of radius $2k^2/\lambda$ cutting the plane PQR, drawn through the point P normal to its radius vector PO, in the circle QR, then the angle OQP satisfies the condition sin OQP $= \lambda\rho/2k^2$, and that AO therefore represents a ray direction which will be reflected by the plane in the original lattice corresponding to the point P, or, as we may say in short, that AO is ray reflected by the point P, the reflected ray being OB. Further, all

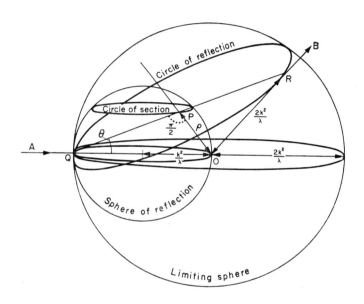

FIG. 3.

rays that can be reflected by P lie along the cone OQR and the incident and reflected rays are diametrically opposite generators of this cone. Here we have a simple geometrical picture of reflection as it occurs in the reciprocal lattice; it may be stated as follows:—

A ray is reflected by a given point of the reciprocal lattice if and only if it lies along a generator of a cone (the Cone of Reflection), whose apex is the origin and whose base is the circle (the Circle of

F

Reflection), which is the intersection of the plane through the point normal to its radius vector, and a sphere (the Limiting Sphere), whose centre is the origin and whose radius is $2k^2/\lambda$. The corresponding reflected ray is the diametrically opposite generator. It is obvious that no real cone of reflection can be formed by any point outside the sphere of radius $2k^2/\lambda$; such points, therefore, cannot reflect, and the sphere is accordingly called the *Limiting Sphere*. The number of reflecting points varies as the volume of this sphere and consequently inversely as λ^3. This is an important consideration in choosing wavelengths.

Now, if instead of finding the possible directions of the incident rays which are reflected by a point P of the reciprocal lattice, we consider the possible positions of the points of the reciprocal lattice which reflects for a given direction of the incident ray AO, we obtain an even more useful construction. It can be seen from Fig. 3 that any point on the sphere QPO satisfies the condition $\sin\theta = \lambda\rho/2k^2$, and this sphere is therefore the locus of all reflecting points for the direction AO of the incident beam. We may state this in the following form:—

A point of the reciprocal lattice reflects a given ray if, and only if, it lies on the surface of a sphere (the Sphere of Reflection), which has the ray for diameter, which passes through the origin and whose radius is k^2/λ.

In the rotation method we may consider the lattice fixed and the incident ray to rotate about it in the equatorial plane carrying the sphere of reflection with it. As each point of the reciprocal lattice passes through the sphere the ray is reflected. In general each point passes through the surface of the sphere twice, giving rise to two reflections, one on either side of the principal plane. If we add to these the times that the inverse point in the lattice passes through the sphere, we have four intersections corresponding to the four symmetrically disposed reflections of a plane in the original lattice. In one complete revolution the sphere of reflection sweeps out a tore. This tore contains all the points which can reflect for any given position of the axis of rotation. If the reflected rays were received on a spherical surface, we should be able to register all the reflections; with the use of plates or cylindrical films we only register those lying in a limited portion of the tore. The sections of the tore registering for plates or films is shown in Fig. 4a. It is plain that one rotation photograph will not contain nearly all the reflecting planes, though it will contain

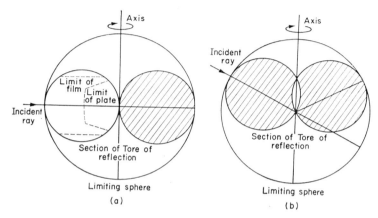

Fig. 4.

the great majority of those near the centre. However, three tores of reflection whose axes are mutually at right angles occupy nearly all the space of the limiting sphere and fill that near the origin at least twice.

6. An Optical Method of Representing the Results of X-ray Analysis*

W. Lawrence Bragg†

1. Introduction

THE representation of the density of scattering matter in a crystal by means of a Fourier series was first suggested by W. H. Bragg.[1] The coefficients of each term in the series can be found by measuring the strength of a corresponding diffracted X-ray beam. The phase of each term is not determined by X-ray measurement, but in the case of simple crystals, and in fact in the case of any centro-symmetrical crystal the structure of which is approximately known, this difficulty can be surmounted. If a crystal has centres of symmetry, all terms in the Fourier series expressing the density have either a maximum or minimum value at these centres; the uncertainty as to the phase of each term reduces to a question of sign alone, and a quite approximate solution of the crystal structure is in general sufficient to determine this sign.

The method of analysis by Fourier series was developed independently by Duane[2] and used by Havighurst[3] and Compton[4] to represent the results of X-ray work. Duane and Havighurst were first in applying it to the much more accurate measurements of X-ray diffraction which had become available

*Zeitschrift für Kristallographie, **70,** 475–92 (1929).
†Professor of Physics, Manchester University.

[1] W. H. Bragg, *Phil. Trans.* A, **215,** 253. (1915)
[2] W. Duane, *Pr. nat. Acad. Washington* **11,** 489 (1925).
[3] R. T. Havighurst, *Pr. nat. Acad. Washington* **11,** 502 (1925).
[4] A. H. Compton, *X-rays and Electrons*, p. 151.

since W. H. Bragg's first suggestion of its usefulness in 1915. The mathematical analysis developed in Ewald's[5] treatment of the reciprocal lattice also leads to the same Fourier representation, though he deals with a series of scattering points and not an extended distribution of scattering matter. It is discussed also by A. L. Patterson.[6]

It will be assumed in the following treatment that one can speak of a "density" ρ of scattering matter in the crystal cell, so that in the element of volume dV at the point x, y, z there is an amount $\rho(x,y,z)dV$ of such matter. The units by which this amount is measured are defined by comparing its scattering of incident radiation with that scattered by a single electron according to classical theory. The quantity $\rho(x,y,z)$ may thus be spoken of as the "electron density" at the point x,y,z, and we may expect that the integration of these density-elements throughout the unit cell will amount to the total number of electrons in the unit cell.

The question as to the soundness of this assumption will not be gone into here; it is sufficient to note that calculations based on it do in actual fact lead to a distribution of scattering matter in the unit cell which outlines atoms, of such size and number as we would expect to be there, and each containing about the right number of electrons.

The Fourier representation, in its most general form, may be summarized as follows.

To each group of indices h, k, l of the crystal structure, there corresponds a complex number $F(hkl)$ which represents the scattering power of the crystalline cell, in the units chosen above, for a diffracted beam whose direction is governed by the indices h, k, l.

$$F(hkl) = \left| F(hkl) \right| e^{i\alpha(hkl)} =$$

$$= \frac{V}{abc} \int_{-a/2}^{a/2} \int_{-b/2}^{b/2} \int_{-c/2}^{c/2} \rho(x,y,z)e^{2\pi i(hx/a + ky/b + lz/c)}dxdydz.$$

[5] P. P. Ewald, *Z. Krist.* **56**, 129 (1921).
[6] A. L. Patterson, *Z. Phys.* **44**, 596 (1927).

V is the total volume of the unit cell, whose axes abc are inclined at any angles to each other, so that $Vdxdydz/abc$ is the element of volume between x and $x+dx$, etc. The origin is taken to be at the centre of the cell.

If now $\rho(x,y,z)$ is represented by a triple Fourier series in x, y, z and this series is substituted in the equations such as (1), a formal solution is given by

$$\rho(x,y,z) = \frac{1}{V}\sum_{-\infty}^{\infty}\sum_{-\infty}^{\infty}\sum_{-\infty}^{\infty} \left| F(hkl) \right| \cos\{2\pi hx/a + 2\pi ky/b + 2\pi lz/c - \alpha(hkl)\}.$$

The summation is made over all positive and negative values of h, k, and l. When $h = k = l = 0$, the corresponding term $F(000)$ is a constant independent of x, y, z, It must be taken to be equal to Z, the total number of electrons in the unit cell of the crystal, for this is necessary in order to satisfy the relation

$$\frac{V}{abc}\int_{-a/2}^{a/2}\int_{-b/2}^{b/2}\int_{-c/2}^{c/2} \rho(x,y,z)dxdydz = Z.$$

The physical interpretation of these equations is very simple. When a diffracted beam is characterized by the integers h, k, l, the matter in the unit cell scatters a wave whose amplitude is equal to that scattered by $\left| F(hkl) \right|$ electrons in the classical theory. This wave has a phase constant $\alpha(hkl)$ when compared with a wave scattered by an electron at the origin. Let us suppose that these amplitudes and phase constants are known for all significant values of h, k, l (when these indices are large the coefficients $F(hkl)$ become very small). The distribution of scattering matter in the unit cell can then be built up. In the first place a uniform distribution Z/V is spread through the cell. To this is added a series of periodic variations in density, crossing each other in all directions. Their amplitudes are given by the coefficients $\left| F(hkl) \right|/V$. If we consider each diffraction as a reflexion by the plane (hkl) of the crystal, the corresponding periodic component is like a train of waves of harmonic form parallel to the plane (hkl). A crest of these waves is situated so that it passes through all points given by

$$2\pi(hx/a+ky/b+lz/c) = \alpha(hkl).$$

This is the feature of X-ray diffraction first pointed out by W. H. Bragg. Each diffracted beam corresponds both in amplitude and phase to that which would be diffracted by a simple harmonic distribution of scattering matter in the crystal, and the total distribution of scattering matter is the sum of all these harmonic fluctuations. W. H. Bragg considered originally the distribution of scattering matter in sheets parallel to one given plane, deriving his coefficients from the different orders of reflexion. The formula above is merely an extension of W. H. Bragg's formula to three dimensions, for a cell of any shape. The formula is expressed somewhat differently to that developed by Duane for the general case, for the present form makes clearer the optical analogy discussed in the next section.

2. *The analogy with image formation in the microscope*

This treatment of X-ray diffraction depends on an optical principle developed fifty years ago by Abbe in discussing the resolving power of the microscope. Abbe's principle was given mathematical form by A. B. Porter,[7] and W. H. Bragg based his work on this paper of Porter's. The principle is illustrated in Fig. 1.

In Fig. 1, $O_1 O_2 O_3$ represent lines of a grating whose rulings are perpendicular to the plane of the diagram. A train of mono-chromatic plane waves falls perpendicularly on this grating. The waves passing through any line, such as O_1, spread in all directions and are received by the lens of a microscope objective represented by a simple line in the figure. After passing through the lens they converge and form an image I_1, of the line O_1.

---

[7]A. B. Porter, *Phil. Mag.* **2,** 154 (1906).

The formation of this image may be considered as taking place in two stages. In the first stage, the waves scattered by all the lines $O_1 O_2 O_3$ etc. recombine to form parallel trains, the "spectra" due to the grating. The lens focusses these spectra at points $S_2 S_1 S_0 S'_1 S'_2$ etc. These may now be considered as sources of monochromatic wave trains, which spread out and form interference fringes in the image space as shown in the figure. These figures compose the image of the grating.

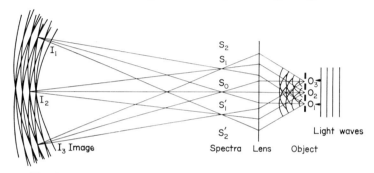

FIG. 1. Abbe's treatment of image formation in the microscope.

The figure shows the interference pattern due to the spectrum of zero order S_0 and the two spectra of first order S_1 and S'_1 alone. For simplicity it may be taken that the grating is symmetrical about the point O_2. The amplitudes of the waves in the image space coming from the points $S_1 S_0 S'_1$ will be taken to be $s_1 s_0 s'_1$ and in this case $s_1 = s'_1$. The amplitude of the light waves in the image space will be clear from the diagram. If a is the distance between I_2 and I_1, and x the distance from I_2 of the point considered, the amplitude then will be

$$s_0 + 2s_1 \cos(2\pi x/a).$$

If all spectra are taken into consideration, the amplitude will be

$$s = s_0 + 2s_1 \cos(2\pi x/a) + 2s_2 \cos(4\pi x/a) + \ldots$$
$$= \sum_{-\infty}^{\infty} s_k \cos(2\pi k x/a).$$

In this case the amplitude of the light in the image space will map out a perfect representation of the amplitude of the light immediately after passing through the grating. This can easily be seen by reversing the light from S_1 S_0 S'_1 etc., and making it pass back through the objective, when it builds up the amplitude everywhere of the light which passes through the grating. The whole diagram is in fact symmetrical about the plane of the spectra, the only distinction between object and image being one of scale. Abbe's treatment of resolving power is based upon the consideration of the extent to which the image will reproduce the object, when a limited number only of the spectra are transmitted by the objective. The greater the number of spectra, the more clearly can we see in the microscope the precise form of the lines of the grating.

Suppose the series of lines O_1 O_2 O_3 to be given a small movement in a downward direction; the spectra S_1 S_0 S'_1 remain fixed in position. There must be, however, a movement upwards of the images I_1, I_2, I_3 and the way in which this is effected is shown by the diagram. Although S_1 remains in the same position, the waves coming from it will be retarded in phase owing to the movement of the grating, and those coming from S'_1 will be advanced. Hence the maxima of the interference fringes all move upwards. The form of the image depends both on the phase and amplitude of the waves coming from the spectra.

Let the simple grating now be replaced by a two dimensional grating (such as the pattern of markings on the surface of a diatom). The spectra now form a doubly infinite series of points, in rows and columns. Each pair of points sends out wave trains which interfere in the image space. The interference fringes are at right angles to the line joining the two spectra, and their spacing is inversely proportional to the distance between the spectra. Each pair of spectra must now be labelled by two indices h, k and the total amplitude of vibration in the image space is given by the formula

$$s = \sum_{-\infty}^{\infty} \sum_{-\infty}^{\infty} 2s(h,k) \cdot \cos 2\pi(hx/a + ky/b).$$

It has again been assumed for the sake of simplicity that there is a centre of symmetry of the pattern on the axis of the instrument. $s(h,k)$ is the amplitude of the wave in the image space due to the spectrum in the column h and row k, and $s(h,k) = s(\bar{h},\bar{k})$.

As A. B. Porter expresses it (*Phil. Mag. loc. cit.* p. 156), a double process of Fourier analysis takes place when light falls on a grating. A simple harmonic grating, if such could be realized, would only give spectra of the first order. If a complex light train falls on it, the light is resolved into a series of harmonic vibrations spread out into a spectrum, so that the grating gives a Fourier analysis of the light. On the other hand, if a plane monochromatic wave train falls on a complex grating a series of spectra are formed, and the amplitude of the light waves in these spectra are proportional to the coefficients of a Fourier series which, when summed, reproduces the form of the grating. In this case the light performs a Fourier analysis of the grating.

In a precisely similar way we use a single periodicity in a crystal to analyse a complex beam of X-rays (analysis of X-ray spectra) and a single X-ray wavelength to analyse a complex periodicity in the crystal (crystal analysis).

The above discussion merely summarizes the optical principle due to Abbe, and the application of Fourier series to X-ray analysis by W. H. Bragg, Ewald, Duane and Compton, and it introduces no new points. In the present paper, I wish to draw a somewhat closer analogy between the examination of a two-dimensional pattern under the microscope, and the examination of all faces round a zone in X-ray analysis of crystals. In Fig. 1, the light falling on the object is first converted into a series of interference maxima, the spectra S_1 S_0 S'_1 etc. These act as new light sources and send out waves which interfere in the image space and produce the image. If there is a large number of spectra, the image will be a faithful representation of the object. All data necessary to build up the image must be available if we know the amplitudes and phases of the waves coming from the spectra, for these are their only characteristics. We may therefore consider the complex system of spectra in the S space as a sort of code message, containing complete information about the form of the object: it is automatically decoded into an image by optical interference in the microscope.

We have a very similar case in X-ray analysis, for we have a table of spectra obtained by experimental measurement. The main difference, as has been often stressed, is that we can only measure intensities and so all information about phase is lost. Further, since we are dealing with a three-dimensional grating, there is no way of forming experimentally the complete series of spectra simultaneously, as can be done with light and a two-dimensional grating. It can be done formally, and in fact this is represented by Ewald's "Reciprocal Lattice". In the reciprocal lattice there is a three-dimensional array of points. Each is defined by three indices h, k, l and corresponds to the diffracted beam characterized by the same indices. The line joining the point to the origin of the reciprocal lattice is perpendicular to the plane (hkl) of the crystal, and its length is inversely proportional to the spacing of planes parallel to (hkl). There is thus a precise analogy between the points of Ewald's reciprocal lattice and the series of spectra S in Fig. 1. Each spectrum $S(h,k)$ is at right angles to the line (hk) in the two-dimensional grating, and its distance from the origin is inversely proportional to the spacing of the lines parallel to (hk). If there were some experimental way of forming simultaneously all possible crystal spectra by using a fourth dimension, they would be arranged on the reciprocal lattice. Their positions, amplitudes, and phases would again contain all information necessary to reform a perfect image of the crystal structure. We may associate with each point of the reciprocal lattice a corresponding amplitude $F(hkl)$. At the origin the amplitude $F(000)$ is equal to Z, the total number of electrons in the unit cell. This is the spectrum of "zero order", the amount of radiation scattered in the direction of the incident rays.

Now the case of X-ray diffraction can be reduced to one of two dimensions by considering only the diffraction of planes around one given zone. We may, for convenience, call the zone the a axis of the crystal, and choose suitable axes b and c to outline a unit crystal cell. The amplitudes and phases of all spectra $F(0kl)$ are supposed to be known. If the crystal is centro-symmetrical $F(0kl)$ will be real, but may be positive or negative. The scattering matter in the crystal is now supposed to be projected on the $b\,c$ plane, when it forms a two-dimensional grating with periodicities b and c along these axes. The density of scattering matter per unit area of the projection, when a single layer of unit cells is projected on the face (100), is given by the formula

$$\rho(y,z) = \frac{1}{A} \left\{ \sum_{-\infty}^{\infty} \sum_{-\infty}^{\infty} F(0kl) \cos{(2\pi ky/b + 2\pi lz/c)} \right\} \quad {}^{(8)}$$

corresponding to the formula for the amplitude in the image in the microscope.

$$s = \sum_{-\infty}^{\infty} \sum_{-\infty}^{\infty} s(kl) \cos{(2\pi ky/b + 2\pi lz/c)}$$

To make the analogy more complete, we may suppose that a two-dimensional grating is constructed, with transparent portions in its pattern corresponding to the atoms in the crystal cells projected on the $b\,c$ plane, but with a scale convenient for optical interference. The spectra of this grating are observed one by one, setting the grating in each case in such a position that incident and diffracted beams make equal angles with the zone [100]. Interference now takes place in exactly the same way both in crystal and in grating, for the phase of the wave scattered by an atom in the crystal only depends on the position of its projection on the face (100).

If these spectra could be replaced by a series of light sources arranged in a corresponding way, and these made to send out waves of the right amplitude and phase, such waves would build up an image of the crystal projected on the $b\,c$ plane with the atoms in correct position and of the correct density. The optical representation of the results of X-ray analysis described here is based on this principle.

---

[8] A is the area of the face (100) of the unit cell.

3. *The optical representation of the projected crystal cell*

Each equal pair of spectra $F(0kl)$ and $F(0\bar{k}\bar{l})$ produces a periodic variation of density in the image. Its amplitude is equal to $2F(0kl)$ and its crests are on the successive $(0kl)$ planes. It is therefore possible to build up an image of the crystal structure if some means can be devised of adding together periodic variations crossing the unit cell in all directions, whose positions and amplitudes correspond to the interference fringes produced by the spectra. The experimental arrangement employed to effect this was as follows. A photographic plate was prepared with a series of light and dark bands on it, the amount of light passing through the plate having as closely as possible a harmonic alternation between the transparent and opaque bands. Actually this was achieved by photographing a row of opaque rods, placed at a distance apart of twice their diameters, illuminated from behind and thrown out of focus. By trial we succeeded in getting a photograph which showed the correct alternation of light and shade (Plate I).

This plate was placed in a projection camera and illuminated from behind, and the shadows of the bands thrown on a screen. The distance of the camera from the screen could be varied and the shadows thrown in any position. A large sheet of photographic paper was pinned on the screen. A hinged flap shielded it from the light when necessary. On this flap an outline of the face $b\ c$ of the crystal cell was drawn to scale.

The bands due to each pair of spectra $(k,l)\ (\bar{k}\bar{l})$ were now thrown in turn on the same sheet of paper. The camera was adjusted until these bands crossed the unit cell on the flap with correct orientation and position (see Plate I). The shutter of the camera was closed, the flap lifted, and then the shutter opened for a time proportional to $F(0kl)$. The flap was then replaced, and the process repeated for the next pair of spectra. In the examples given here, about forty such sets of bands were thrown on each sheet of photographic paper. The paper was then developed, and the result is a projection of the unit cell with all its atoms in their correct positions (Plates II, III and IV).

This method does not reproduce exactly the formation of an optical image. By throwing on the paper the light and dark bands, one is adding an amount of light proportional to

$$F(0kl)\{1+\cos{(2\pi ky/b+2\pi lz/c)}\}.$$

The dark bands should contribute a negative component to the light which falls on the paper, but as this cannot be realized experimentally, the final image represents the summation of the Fourier series with in addition a large constant component. This does not, however, prevent our obtaining a very realistic "microphotograph" of the crystal structure. By adjusting the scale of times of exposure and using the properties of response to illumination of the paper, one can ensure that this constant component does not fog the background so much as to prevent the atoms in the crystal from showing clearly.

In calculating the times for which the shutter must be opened, allowance must be made for the movement of the camera towards and away from the screen, since this alters the intensity of illumination in the bright bands. This allowance was made by opening the shutter for a time proportional to $F(0kl)\cdot r^2$, where r is the distance from camera to screen.

4. *Projections of diopside structure*

The figures illustrate the application of this method to the crystal Diopside, $CaMg(SiO_3)_2$. In a recent paper in this journal, Mr. Warren and the author[9] used a series of F values around the [100], [010], and [001] zones to fix the atomic positions. These F values are shown in Figs. 2a, 2b, 2c. They are spaced so as to represent the cross-spectra given by the corresponding projections as two-dimensional gratings. Each spectrum is shown as a number. It is positive or negative, according to whether the diffracted wave has a phase the same as, or opposite to, a wave scattered by an electron at the origin in the crystal cell. The number represents the value of $F(hkl)$. These sets of figures constitute central

9B. Warren and W. L. Bragg, *Z. Krist.* **69,** 167 (1928).

sections of Ewald's reciprocal lattice, taken through the origin. Each section is perpendicular to the corresponding crystal zone.

In the following figures are shown

a. The positions of the atoms in the crystal cell, as determined by Mr. Warren and the author by X-ray analysis of the usual type. These positions are determined by fourteen parameters, all of which we evaluated by using quantitative measurements of X-ray diffraction (Figs. 3(a), 4(a), 5(a)).

b. The summation of the Fourier series by calculation. The results of these calculations are given in a paper recently published in the *Proceedings of the Royal Society*[10]. The Fourier series is summed for a number of points over each face of the unit cell. The result is a set of numbers giving the density of scattering matter in the cell as projected on that face. In the diagram, contour lines are drawn through points of equal density like the contour lines on a map, so that points of high density are surrounded by rings closing in to a "peak". The comparison of the peaks shown in the Fourier distribution with the atomic positions determined in the usual way is very interesting. The agreement is remarkably good (Figs. 3(b), 4(b), 5(b)).[11]

c. The optical projection of the crystal cell. *This is far from perfect, the chief cause of error being the difficulty of getting a harmonic distribution of light intensity in the projected bands. Nevertheless the atoms show up in a most realistic way, just as if one were obtaining a "microphotograph" of the crystal structure. It is interesting to note that there are, in the projection, optical defects just like those which one gets in a microscope when the object is near the limit of the resolving power. In the projection on the face (001)†, for instance, there is in a certain position an optical "ghost" (the position is marked *P* in the corresponding Fig. 5a of the crystal structure). This defect is not due to the

---

[10]W. L. Bragg, *Proc. Roy. Soc.* A, **123**, 537 (1929).

[11]In the Royal Society paper it is shown that the fourteen parameters *u, v, w* determined by Fourier analysis agree with those previously assigned to the crystal, with an average discrepancy of 0·5%.

*Plates II, III, IV.   †Plate IV.

method of projection, for it also appears in the summation of the Fourier series. It corresponds precisely to defects due to lack of resolving power, the total number of spectra measured being

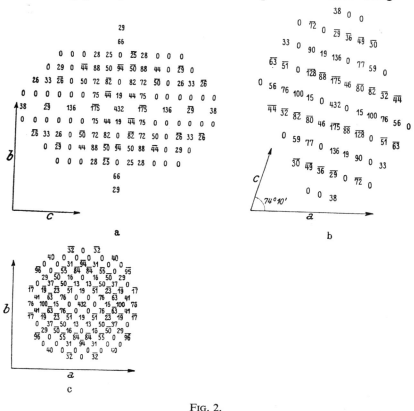

FIG. 2.

FIG. 2a. Spectra around zone [100], Diopside.
FIG. 2b. Spectra around zone [010], Diopside.
FIG. 2c. Spectra around zone [001], Diopside.

insufficient to give a perfect image of the structure. In the Fourier projection one can count the number of electrons associated with each peak and so recognise what atoms they correspond to.

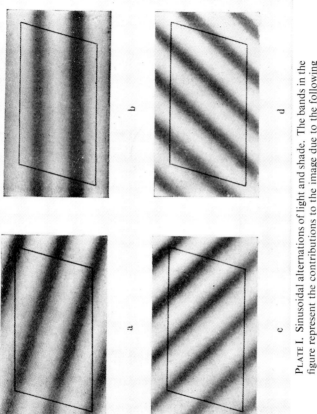

PLATE I. Sinusoidal alternations of light and shade. The bands in the figure represent the contributions to the image due to the following spectra (a) $F(102)$, phase negative. (b) $F(002)$, phase positive. (c) $F(302)$, phase negative. (d) $F(30\bar{1})$, phase positive.

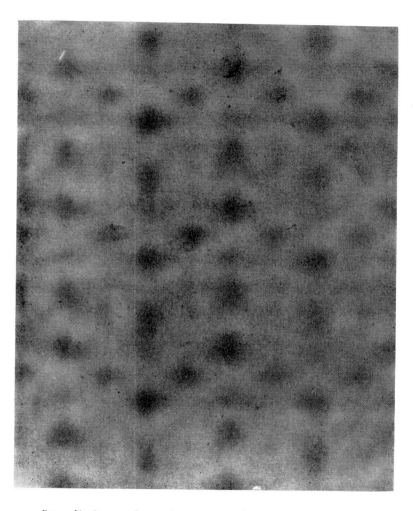

PLATE II. Image of crystal structure projected on face (100). The image is formed by superposition of bands such as those shown in Plate I, the time of exposure to each band being proportional to the numbers in Fig. 2a.

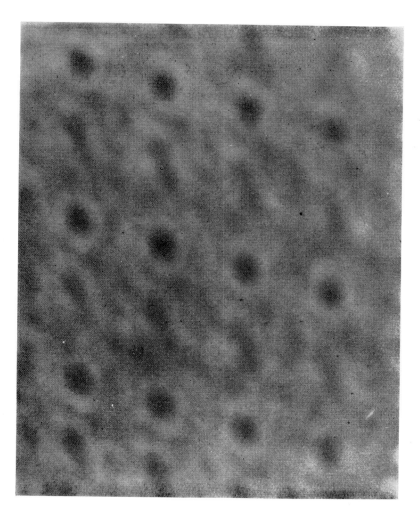

Plate III. Image of crystal structure, projection on face (010).

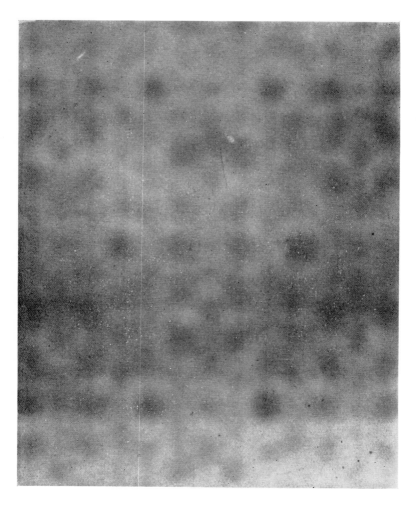

PLATE IV. Image of crystal structure, projection on face (001).

This particular small peak only contains 1·5 electrons and is clearly due to the lack of sufficient terms in the Fourier series.

_____

It is with great pleasure that I acknowledge my indebtedness to my father, Sir William Bragg, for suggestions which materially

Fig. 3a.   Fig. 3b.

Fig. 3a. Atomic positions, projection on face (100).
Fig. 3b. Summation of Fourier series for projection on face (100).
The distribution of scattering matter is indicated by contour lines
drawn through points of equal density in the projection.

contributed to the work described in this paper. A few months ago we discussed the possibility of making a more satisfactory use of the Fourier method. We had approached this problem along slightly different lines. He had attempted to apply two-dimensional and three-dimensional Fourier series to an organic substance, diphenyl, and had found that the first few terms of the Fourier series did in fact outline the general massing of scattering matter in the unit cell if certain assumptions about phase were made. I had been interested in the analogy between X-ray analysis and optical image formation, and the general relation of Fourier analysis to analysis by trial and error, as outlined briefly in

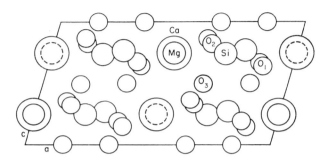

FIG. 4a. Atomic positions, projection on face (010).

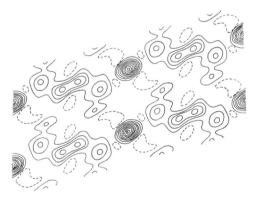

FIG. 4b. Summation of Fourier series, projection on face (010).

a recent paper in the Zeitschrift by Mr. West and myself. We had, in our extensive measurements on inorganic crystals, sufficient information to make very faithful projections of the crystal by using Fourier series, and my father's results led me to take up the problem of computing the series for all planes around a crystal zone. The computation, which had at first seemed to be very lengthy, turned out to be surprisingly rapid in practice, and the resulting projections proved to be remarkable images of the scattering matter in the crystal.

I have also to thank Mr. J. West for his kind help in the preparation of this paper, and in particular for the taking of the photographs which illustrate it.

Summary

The paper describes an optical method of reproducing the results of X-ray analysis. Use is made of the analogy between the principle underlying the interpretation of X-ray results, and the principle formulated by Abbe in his treatment of the resolving power of the microscope (illustrated in Fig. 1). The microscope is supposed to be focussed on a two-dimensional grating illuminated by parallel monochromatic light. A series of spectra are formed behind the objective, and trains of light waves from these spectra interfere so as to build up the image. In the case of X-ray analysis, one can measure a series of diffracted beams reflected by planes around a given zone. These correspond to the spectra formed by a two-dimensional grating which is the projection in a direction parallel to that zone of the scattering matter in the crystal cell on a plane. In the microscope, every pair of opposite spectra contributes a single series of interference bands to the image. The X-ray spectra are now treated as if they were optical spectra, each pair of X-ray measurements around a zone [100], such as $F(0kl)$ and $F(0\bar{k}\bar{l})$ contributing a periodic fluctuation of density to the projection on the face (100) of the crystal.

A sheet of photographic paper is pinned on a screen, with the form of the face (100) of the crystal cell outlined upon it. A plate is prepared with a series of harmonic bands of light and shade, like interference fringes, upon it. This is placed in a projection camera and illuminated from behind, and the bands are thrown upon the paper. The camera is adjusted for each pair of spectra

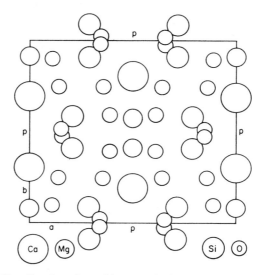

FIG. 5a. Atomic positions, projection on face (001).

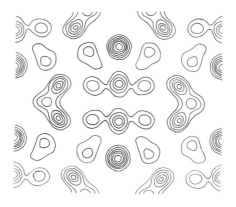

FIG. 5b. Summation of Fourier series, projection on face (001).

$F(0kl)$, $F(0\bar{k}\bar{l})$ so that the bands are in the correct position, and the shutter of the camera is opened for a time corresponding to the amplitude of the diffracted beam of X-rays as measured experimentally. About forty such bands are projected in turn on the same sheet of paper. It is necessary to have a previous approximate analysis of the crystal structure in order to know whether a light or a dark band passes through the centre of the cell, which is a centre of symmetry of the crystal structure.

When the paper is developed it shows an image of the crystal cell as projected on one of its faces, all the atoms appearing in their correct positions. By projecting on the three faces of the cell in turn, the positions of the atoms in space can be found. These "microphotographs" of the crystal structure are shown in Plates II, III, IV, the crystal being in this case Diopside, $CaMg(SiO_3)_2$. The majority of the fourteen parameters which determine the positions of the atoms in the crystal can be fixed quite accurately from the photographs.

Received April 3rd 1929.

7.   An Improved Numerical Method of Two-dimensional Fourier Synthesis for Crystals*

H. LIPSON and C. A. BEEVERS†

Abstract

It is shown that a two-dimensional Fourier summation for a crystal without a centre of symmetry can be resolved into one-dimensional summations, and these latter can be calculated very rapidly by using a set of printed strips which give cosine and sine waves of different wavelengths and amplitudes. The most useful interval of division of the strips, and various features concerning their use, are described.

§ 1. *Introduction*

A method has been described by J. M. Robertson[1] whereby the summation of a two-dimensional Fourier series is reduced to the processes of first arranging sets of figures, and of then adding them.   The method described in the present paper is similar to this, but the process of arrangement is very greatly reduced, and since most of the time taken is concerned with the additions it is claimed that the method is as quick as any numerical method can possibly be.

The method is based on that described by the authors.[2] Essentially this reduces the computations to the addition of a

---

Proceedings of the Physical Society, **48,** 772–80 (1936).

†Physical Laboratories, University of Manchester. Communicated by Prof. W. L. Bragg, April 24, 1936.   Read in title June 5, 1936.

[1] Robertson, J. M. *Phil. Mag.* **21,** 176 (1936).   We are indebted to the author for allowing us to see his manuscript before publication.

[2] Beevers, C. A. and Lipson, H. *Phil. Mag.* **17,** 855 (1934).

number of one-dimensional cosine and sine terms. There are thus only two variables, the amplitude and wavelength of each cosine or sine wave, and the number of combinations of these to cover most cases is sufficiently small for it to be practicable to prepare them in advance. This has been done by the authors, the numbers representing the various waves being printed on cardboard strips. The present paper describes the derivation and use of such a set of strips.

§ 2. *Theoretical Considerations*

Division of the area of the projection. It has been shown[3] that if a radiation of wavelength λ be used, and if all reflections with glancing angles less than θ be observed, then Fourier synthesis will not show any detail finer than $0.6\lambda/2 \sin \theta$. If $\lambda = 1.54$ Å. (for $K\alpha$ radiation of copper) and $\sin \theta = 1$, this quantity is 0.46 Å. Thus it is necessary that the points at which the electron-density is computed should be appreciably closer together than this; but any undue closeness will be unnecessary as it will not reveal any greater detail.

Thus for most ordinary projections, the edges[4] of which are between 8 and 15 Å., a division into about 50[5] parts would be suitable. This number, however, has some disadvantages. For instance points with co-ordinates $\frac{1}{4}$ are not included explicitly, and it is sometimes important that the electron-density at such points should be known accurately. This defect may be overcome by dividing the edge into 48[6] or 60[7] parts, and this also makes the work of preparation of the tables rather lighter. The latter number was chosen because it extends the limit of applicability of the method somewhat.

---

[3] Bragg, W. L. and West, J. *Phil. Mag.* **10**, 823 (1930).
[4] These will not be the edges of the unit cell unless the crystallographic axes are orthogonal.
[5] Bragg, W. L. *Proc. Roy. Soc.* A, **123**, 537 (1929).
[6] Patterson, A. L. *Z. Kristallogr.* (A) **90**, 517 (1935).
[7] See Note 1.

Adaptibility to extreme cases. The division of the cell-edges into 60 parts is directly suitable if the maximum indices are each between 11 and 20. If one of the maximum index-values is below 11 (i.e. if one cell-dimension is small) it will be advisable to make the interval of division of the corresponding edge $\frac{1}{30}$ This may be done most simply by considering that we are going to perform a synthesis of two unit cells, the short edge being considered as doubled. One of the indices of each plane referred to this doubled cell will be doubled. Thus all we have to do in order to obtain an interval of $\frac{1}{30}$ in any direction is to double the corresponding indices of the Fs and proceed in the usual way. The strips will then give the projection over the range 0 to 0·50 directly, instead of from 0 to 0·25. Similarly, if the maximum index in one direction is 6, the indices corresponding to this direction may be trebled, and the interval of division will be $\frac{1}{20}$ The general rule is that the tables should be used as near as possible to their maximum index 20.

If, on the other hand, the maximum index exceeds 20 the strips as prepared by the authors will not be comprehensive. It will be desirable to make the interval of division $\frac{1}{120}$, in which case the indices must be halved. Thus the odd indices cannot be included, and their contribution will have to be computed by the ordinary method. Nevertheless, unless both ranges of index are greater than 20, it will be possible to use the strips directly in the final tables which involve the greater part of the work.

Expression of the general case in terms of one-dimensional series. The most general case (so long as the wavelength used is not near an absorption edge of any of the atoms concerned) is that of the completely asymmetric projection, in which the only relation between the Fs is $F(hk0) = F^*(\bar{h}\bar{k}0)$, where F^* is the conjugate complex of F. The direction of projection is taken as parallel to the c axis.

If x and y are the co-ordinates of any point in the projection in fractions of 2π, and H and K are the maximum values of h and k in the array of Fs, the summation required is as follows, $F(hk)$

being now written for $F(hk0)$

$$\sum_{-H}^{H} \sum_{-K}^{K} F(hk) \exp i (hx+ky)$$

$$= \sum_{-H}^{H} \sum_{1}^{K} \{F(hk) \exp i (hx+ky) + F^*(hk) \exp -i (hx+ky)\}$$

$$+ \sum_{1}^{H} \{F(h0) \exp ihx + F^*(h0) \exp -ihx\}$$

$$+ F(00).$$

Since $F(hk)$ and $F(\bar{h}\bar{k})$ are conjugate, they may be expressed respectively as $|F(hk)| \exp i\delta(hk)$ and $|F(hk)| \exp -i\delta(hk)$, and the expression becomes,

$$\sum_{-H}^{H} \sum_{1}^{K} |F(hk)| \{\exp i [hx+ky+\delta(hk)] + \exp -i [hx+ky+\delta(hk)]\}$$

$$\sum_{1}^{H} |F(h0)| \{\exp i [hx+\delta(h0)] + \exp -i [hx+\delta(h0)]\}$$

$$+ F(00)$$

$$= 2 \sum_{-H}^{H} \sum_{1}^{K} |F(hk)| \cos [hx+ky+\delta(hk)] + 2 \sum_{1}^{H} |F(h0)| \times$$
$$\times \cos [hx+\delta(h0)]$$

$$+ F(00).$$

Let $\qquad |F(hk)| \cos \delta(hk) = C(hk)$

and $\qquad |F(hk)| \sin \delta(hk) = S(hk).$

Then we have

$$2 \sum_{-H}^{H} \sum_{1}^{K} \{C(hk) [\cos hx \cos ky - \sin hx \sin ky]$$
$$- S(hk) [\sin hx \cos ky + \cos hx \sin ky]\}$$

$$+ 2 \sum_{1}^{H} \{C(h0) \cos hx - S(h0) \sin hx\}$$

$$+ F(00)$$

which, for the purposes of summation, is put into the form

$$2 \sum_1^H \sum_1^K \{C(hk) [\cos hx \cos ky - \sin hx \sin ky]$$
$$- S(hk) [\sin hx \cos ky + \cos hx \sin ky]$$
$$+ C(\bar{h}k) [\cos hx \cos ky + \sin hx \sin ky]$$
$$(\bar{h} - Sk) [-\sin hx \cos ky + \cos hx \sin ky]\}$$

$$+ 2 \sum_1^K \{C(0k) \cos ky - S(0k) \sin ky\}$$

$$+ 2 \sum_1^H \{C(h0) \cos hx - S(h0) \sin hx\}$$

$$+ F(00)$$

$$= 2 \sum_1^H \sum_1^K \{[C(hk) + C(\bar{h}k)] \cos hx \cos ky$$
$$- [C(hk) - C(\bar{h}k)] \sin hx \sin ky$$
$$- [S(hk) - S(\bar{h}k)] \sin hx \cos ky$$
$$- [S(hk) + S(\bar{h}k)] \cos hx \sin ky\}$$

$$+ 2 \sum_1^K \{C(0k) \cos ky - S(0k) \sin ky\}$$

$$+ 2 \sum_1^H \{C(h0) \cos hx - S(h0) \sin hx\}$$

$$+ F(00).$$

The preliminary tables would then be concerned with the evaluation of

$$2 \sum_1^H \{C(h0) \cos hx - S(h0) \sin hx\} + F(00) = A(0, x), \text{ say,}$$

of

$$2 \sum_1^H \{[C(hk) + C(\bar{h}k)] \cos hx - [S(hk) - S(\bar{h}k)] \sin hx\}$$
$$+ 2C(0k) = A(k, x),$$

and of

$$2 \sum_{1}^{H} \{[C(hk) - C(\bar{h}k)] \sin hx + [S(hk) + S(\bar{h}k)] \cos hx\} + 2S(0k)$$
$$= B(k, x).$$

$A(k, x)$ and $B(k, x)$ are evaluated separately for each value of k from 1 to K.

The final summation is then given by

$$A(0, x) + \sum_{1}^{K} A(k, x) \cos ky - \sum_{1}^{K} B(k, x) \sin ky.$$

Thus even the general expression can be reduced to the summation of one-dimensional series. In practice, however, this case is not important, as the values of $C(hk)$ and $S(hk)$ are not determinable by experiment. If, however, the projection has a centre of symmetry at the origin (which may be produced by a two-fold axis in the structure perpendicular to the plane of projection)

$$F(hk) = F(\bar{h}\bar{k})$$

and thus the Fs are real.

$$\therefore C(hk) = F(hk) \text{ and } S(hk) = 0.$$

In this case

$$A(0, x) = 2 \sum_{1}^{H} \{F(h0) \cos hx\} + F(00),$$

$$A(k, x) = 2 \sum_{1}^{H} \{[F(hk) + F(\bar{h}k)] \cos hx\} + 2F(0k),$$

and $\quad B(k, x) = 2 \sum_{1}^{H} \{[F(hk) - F(\bar{h}k)] \sin hx\}.$

The projection may have other symmetries which involve relationships between $F(hk)$ and $F(\bar{h}k)$. In such cases the summation is even simpler.

§ 3. *Practical Details*

Derivation of numbers. The complete equipment as prepared by the authors consists of 2079 cosine strips and 1980 sine strips. The cosine strips have index-values 0 to 20 inclusive, and the sine strips 1 to 20 inclusive. For each index we have strips of amplitudes $\overline{99}$ to 99. The functions are plotted at intervals of 6° so that the strips corresponding to $A \sin hx$, for example, has printed on it the value of $A \sin (2\pi nh/60)$, correct to the nearest whole number, where n has values from 0 to 15. Each strip has the corresponding negative function on the back. As an example the sine strip for index 3 and amplitude 99 is:

99 S3	0	31	58	80	94	99	94	80	58	31	0	$\overline{31}$	$\overline{58}$	$\overline{80}$	$\overline{94}$	$\overline{99}$

The strips are cut from a number of tables, each table having one index of cosine or sine with the amplitudes in successive rows.

The different tables are printed from the same columns of type merely arranged in different ways. For example, if the columns for sine 1 are numbered 0, 1, 2, 3, 4, 5, 6, 7, 8, 9, 10, 11, 12, 13, 14, 15, then for sine 2 we have the arrangement of columns 0, 2, 4, 6, 8, 10, 12, 14, 14, 12, 10, 8, 6, 4, 2, 0; and for cosine 6 we have the order: 15, 9, 3, $\overline{3}$, $\overline{9}$, $\overline{15}$, $\overline{9}$, $\overline{3}$, 3, 9, 15, 9, 3, $\overline{3}$, $\overline{9}$, $\overline{15}$, the superior negative signs indicating that all the numbers in the column are negative instead of positive.

The area of summation. If we deal with crystals for which the projection possesses a centre of symmetry it is necessary to obtain the projection for only half of the unit cell. Sometimes the projection will have other symmetry elements which will still further reduce the area to be evaluated. The largest area which will have to be dealt with is given by the conditions that one parameter must go from 0 to 0·50 and the other from 0 to 1·00. This area can, however, easily be derived from the area 0 to 0·25 by 0 to 0·25 which is the area covered directly by the strips. The general rule which enables this extension of area to be effected is that in carrying out any of the one-dimensional summations even indices and odd indices must be kept separate. If, for example, we are summing cosine terms, we see that cos hx is symmetrical about $x = 0.25$ when h is even, and is antisymmetrical about $x = 0.25$ when h is odd. Thus, if in the summation we keep separate the terms with h even and h odd, we have only to add and subtract the two sets of totals in order to obtain the sums for $x = 0$ to 0.25 and $x = 0.50$ to 0.25, respectively. In the case of sine terms we sum first for the planes with h odd, and then for those with h even. Sum and difference then give $x = 0$ to 0.25 and $x = 0.50$ to 0.25, respectively. These ways of extending the range are illustrated in Table III, where they are applied to the summations of cos ky and sin ky.

To extend the range of a parameter to 0.50 to 1.00 we notice that all the cosine terms are symmetrical about 0.50 whilst all the sine terms are antisymmetrical about this point. We therefore have to subtract sine terms from cosine terms to give the range 0 to 0.50 and to add them to give the range 1.00 to 0.50. In this way we derive the whole of the projection. The process is illustrated in Table IV.

Storage of strips. The strips are kept in two similar wooden boxes[8], one for cosine strips and one for sines. All the strips of one index are kept in one compartment, the partitions between the compartments being made of aluminium. The sloping sides of the boxes ensure that on removal of one strip from a compartment its place shall be left open for its reinsertion. If desired the boxes may be made portable by fitting them with lids, some device being used to clamp the strips.

§ 4. *Precautions against errors*

Mistakes in the use of the strips can arise from selection of the wrong strip or from errors in addition of the numbers. The

[8]Beevers, C. A. and Lipson, H. *Nature* **137**, 825 (1936).

former can be guarded against by a simple check. The strips are removed from the boxes by selecting the index, and then the amplitude required for that index, and so on, the order of consideration being thus index, amplitude, index, amplitude, and so on. They should be checked by a glance down the index column, which will in general be 1, 3, 5, 7, . . . or 0, 2, 4, 6, . . ., and then by looking down the amplitude column, with special attention to signs. Mistakes in adding will be guarded against by the repetition of each addition. Especial care should be taken

TABLE I. VALUES OF $F(hk0)$
$F(hk0) = F(\bar{h}\bar{k}0)$

k	$\bar{8}$	$\bar{7}$	$\bar{6}$	$\bar{5}$	$\bar{4}$	$\bar{3}$	$\bar{2}$	$\bar{1}$	0	1	2	3	4	5	6	7	8
0	8	11	12	0	28	0	33	26	129	26	33	0	28	0	12	11	8
1		11	0	20	0	21	$\overline{18}$	30	0	18	10	19	$\overline{10}$	0	0	0	
2			0	0	0	0	$\overline{21}$	0	$\overline{17}$	12	$\overline{32}$	17	$\overline{7}$	10	0	9	0
3			0	0	18	7	26	$\overline{20}$	23	11	50	17	18	7	16	6	8
4		0	18	13	26	8	27	17	0	0	17	10	8	0	10		
5		0	0	0	0	12	0	0	$\overline{14}$	11	$\overline{15}$	0	$\overline{10}$	0	0		
6		0	0	0	7	0	16	0	12	14	15	0	9	0	0		
7		7	0	13	9	19	15	21	15	18	0	19	7	15			
8		0	10	0	0	0	16	$\bar{8}$	12	0	11	0	0	0			
9			0	0	0	0	$\overline{11}$	0	0	0	0	7	0				
10			8	0	12	0	14	7	16	0	10	0	12				
11				13	0	12	0	17	0	8	0	7					
12				0	0	0	0	0	0	0	0						
13					0	0	0	11	0	7							

over the preliminary tables as an error here will run through a large part of the work.

If desired a complete check could be made by carrying out the summation the other way round. If for example the synthesis was done by first collecting the Fs with the same h in the preliminary tables and multiplying by the cos ky functions, it could be repeated by collecting first the Fs with the same k and multiplying by the cos hx functions. Usually, however, it will be enough to work out one line of the synthesis in this way.

§ 5. *Example of the use of the strips*

In order to compare the procedure with that of the method previously described[2] the same example will be used.

TABLE IIa
VALUES OF $F(hk0) + F(\bar{h}k0)$

k	2h=0	2	4	6	8	10	12	14	16
0	65	26	33	0	28	0	12	11	8
2	12	$\overline{49}$	17	$\overline{28}$	10	0	9	0	
4	0	17	44	18	34	13	28	0	
6	12	14	31	0	16	0	0	0	
8	12	$\overline{8}$	27	0	0	0	10	0	
10	16	7	24	0	24	0	8		
12	0	0	0	0	0	0			
1	0	48	$\overline{8}$	40	$\overline{10}$	20	0	11	
3	11	73	$\overline{3}$	44	14	34	6	8	
5	$\overline{14}$	11	$\overline{15}$	12	$\overline{10}$	0	0	0	
7	15	39	15	38	16	28	0	7	
9	0	0	$\overline{11}$	7	0	0	0		
11	0	25	0	19	0	13			
13	0	18	0	0	0				

TABLE IIb
VALUES OF $F(hk0) - F(\bar{h}k0)$

k	2h=0	2	4	6	8	10	12	14	16
0	0	0	0	0	0	0	0	0	0
2	0	$\overline{15}$	17	14	10	0	0	0	
4	0	$\overline{17}$	$\overline{10}$	2	$\overline{18}$	$\overline{13}$	$\overline{8}$	0	
6	0	14	$\overline{1}$	0	2	0	0	0	
8	0	8	$\overline{5}$	0	0	0	$\overline{10}$	0	
10	0	7	$\overline{4}$	0	0	0	$\overline{8}$		
12	0	0	0	0	0	0			
1	0	$\overline{12}$	28	2	$\overline{10}$	$\overline{20}$	0	$\overline{11}$	
3	0	27	37	8	0	$\overline{2}$	6	8	
5	0	11	$\overline{15}$	$\overline{12}$	$\overline{10}$	0	0	0	
7	0	$\overline{3}$	$\overline{15}$	0	$\overline{2}$	2	0	$\overline{7}$	
9	0	0	11	7	0	0	0		
11	0	$\overline{9}$	0	$\overline{5}$	0	$\overline{13}$			
13	0	4	0	0	0				

Table I shows the values of $F(hk)$ for $CuSO_4 . 5H_2O$, which has only centro-symmetry. It will be observed that the maximum value of h is 8, whereas that of k is 13. It is therefore advisable to sum first with respect to k, so that the largest summations are confined to the shorter preliminary tables. Moreover, since the maximum value of h is only 8, the h indices may be multiplied by 2 in order to double the interval of division along the corresponding edge of the projection.

It is convenient next to prepare Tables IIa and IIb which give the values of $F(hk) + F(\bar{h}k)$ and $F(hk) - F(\bar{h}k)$ respectively. Where one index is zero the values of $F(h0)$ and $F(0k)$ only are used, and the value of $F(00)$ is divided by 2. Comparison with the general formula will show that this is equivalent to dividing

TABLE IIIa

Evaluation of $\frac{1}{2}A(2h, y)$ for $2h = 6$, $= \overset{K}{\underset{1}{\Sigma}}[F(6k) + F(\bar{6}k)] \cos ky + F(60)$

			y (60ths)															
			0	1	2	3	4	5	6	7	8	9	10	11	12	13	14	15
$\overline{28}$	C	2	$\overline{28}$	$\overline{27}$	$\overline{26}$	$\overline{23}$	$\overline{19}$	$\overline{14}$	$\overline{9}$	$\overline{3}$	3	9	14	19	23	26	27	28
18	C	4	18	16	12	6	$\overline{2}$	$\overline{9}$	$\overline{15}$	$\overline{18}$	$\overline{18}$	$\overline{15}$	$\overline{9}$	$\overline{2}$	6	12	16	18
			$\overline{10}$	$\overline{11}$	$\overline{14}$	$\overline{17}$	$\overline{21}$	$\overline{23}$	$\overline{24}$	$\overline{21}$	$\overline{15}$	$\overline{6}$	5	17	29	38	43	46
40	C	1	40	40	39	38	37	35	32	30	27	24	20	16	12	8	4	0
44	C	3	44	42	36	26	14	0	$\overline{14}$	$\overline{26}$	$\overline{36}$	$\overline{42}$	$\overline{44}$	$\overline{42}$	$\overline{36}$	$\overline{26}$	$\overline{14}$	0
12	C	5	12	10	6	0	$\overline{6}$	$\overline{10}$	$\overline{12}$	$\overline{10}$	$\overline{6}$	0	6	10	12	10	6	0
38	C	7	38	28	4	$\overline{22}$	$\overline{37}$	$\overline{33}$	$\overline{12}$	15	35	36	19	$\overline{8}$	$\overline{31}$	$\overline{38}$	$\overline{25}$	0
7	C	9	7	4	$\overline{2}$	$\overline{7}$	$\overline{6}$	0	6	7	2	$\overline{4}$	$\overline{7}$	$\overline{4}$	2	7	6	0
19	C	11	19	8	$\overline{13}$	$\overline{18}$	$\overline{2}$	16	15	$\overline{4}$	$\overline{19}$	$\overline{11}$	9	19	6	$\overline{14}$	$\overline{17}$	0
			160	132	70	17	0	8	15	12	3	3	3	$\overline{9}$	$\overline{35}$	$\overline{53}$	$\overline{40}$	0
Difference			150	121	56	0	$\overline{21}$	$\overline{15}$	$\overline{9}$	$\overline{9}$	$\overline{12}$	$\overline{3}$	8	8	$\overline{6}$	$\overline{15}$	3	
Sum			$\overline{170}$	$\overline{143}$	$\overline{84}$	$\overline{34}$	$\overline{21}$	$\overline{31}$	$\overline{39}$	$\overline{33}$	$\overline{18}$	$\overline{9}$	2	26	64	91	83	46
			30	29	28	27	26	25	24	23	22	21	20	19	18	17	16	15
						y (60ths)												

the whole series by 2, the smaller numbers making the subsequent additions less laborious. This process will be different in cases of higher symmetry for which, for example, $F(hk) = F(\bar{h}k)$, when the series may be divided by 4. Then the numbers in Table IIb will be zero, and in Table IIa will be the single F values, except that $F(h0)$ and $F(0k)$ will be divided by 2, and $F(00)$ by 4. The process is most easily visualized in terms of the reciprocal lattice.[9] To each point (h, k) of the ab section of the lattice is attached a weight $F(hk)$. If $F(hk) = F(\bar{h}k)$ each weight $F(hk)$ occurs four times, $F(0k)$ and $F(h0)$ twice, and $F(00)$ once.

Tables IIIa and IIIb are the parts of the preliminary tables for which $2h = 6$, and in these tables the actual strips used are depicted.

TABLE IIIb

EVALUATION OF $\frac{1}{2}B(2h, y)$ FOR $2h = 6, = \overset{K}{\underset{1}{\Sigma}}[F(6k) - F(\bar{6}k)] \sin ky$

			y (60ths)															
			0	1	2	3	4	5	6	7	8	9	10	11	12	13	14	15
$\bar{2}$	S	1	0	0	0	$\bar{1}$	$\bar{1}$	$\bar{1}$	$\bar{1}$	$\bar{1}$	$\bar{1}$	$\bar{2}$	$\bar{2}$	$\bar{2}$	$\bar{2}$	$\bar{2}$	$\bar{2}$	$\bar{2}$
$\bar{8}$	S	3	0	$\bar{2}$	$\bar{5}$	$\bar{6}$	$\bar{8}$	$\bar{8}$	$\bar{8}$	$\bar{6}$	$\bar{5}$	$\bar{2}$	0	2	5	6	8	8
$\bar{12}$	S	5	0	$\bar{6}$	$\bar{10}$	$\bar{12}$	$\bar{10}$	$\bar{6}$	0	6	10	12	10	6	0	$\bar{6}$	$\bar{10}$	$\bar{12}$
7	S	9	0	6	7	2	$\bar{4}$	$\bar{7}$	$\bar{4}$	2	7	6	0	$\bar{6}$	$\bar{7}$	$\bar{2}$	4	7
$\bar{5}$	S	11	0	$\bar{5}$	$\bar{4}$	2	5	2	$\bar{3}$	$\bar{5}$	$\bar{1}$	4	4	$\bar{1}$	$\bar{5}$	$\bar{3}$	2	5
			0	$\bar{7}$	$\bar{12}$	$\bar{15}$	$\bar{18}$	$\bar{20}$	$\bar{16}$	$\bar{4}$	10	18	12	$\bar{1}$	$\bar{9}$	$\bar{7}$	2	6
14	S	2	0	3	6	8	10	12	13	14	14	13	12	10	8	6	3	0
2	S	4	0	1	1	2	2	2	1	0	0	$\bar{1}$	$\bar{2}$	$\bar{2}$	$\bar{2}$	$\bar{1}$	$\bar{1}$	0
			0	4	7	10	12	14	14	14	14	12	10	8	6	5	2	0
Difference			0	$\bar{3}$	$\bar{5}$	$\bar{5}$	$\bar{6}$	$\bar{6}$	$\bar{2}$	10	24	30	22	7	$\bar{3}$	$\bar{2}$	4	6
Sum			0	$\bar{11}$	$\bar{19}$	$\bar{25}$	$\bar{30}$	$\bar{34}$	$\bar{30}$	$\bar{18}$	$\bar{4}$	6	2	$\bar{9}$	$\bar{15}$	$\bar{12}$	0	6
			30	29	28	27	26	25	24	23	22	21	20	19	18	17	16	15
					y (60ths)													

Of course, only the totals would be written down. The amplitude columns of the strips used in Tables III correspond with columns (shown in italics) in Tables IIa and IIb. The results of Tables IIIa, IIIb are given in the rows designated "difference" and "sum", which give the totals corresponding to the two ranges of y, 0 to 15/60, and 30/60 to 15/60.

[9] Bragg, W. L. *The Crystalline State* 1, 235.

G

TABLE IV. FINAL SUMMATIONS FOR $y = 7/60$

$$A(0, 7/60) + \sum_{1}^{H} A(h, 7/60) \cos hx - \sum_{1}^{H} B(h, 7/60) \sin hx$$

| | \multicolumn{16}{c}{x (30ths)} | | | | | | | | | | | | | | |
	0	1	2	3	4	5	6	7	8	9	10	11	12	13	14	15
93 C 0	93	93	93	93	93	93	93	93	93	93	93	93	93	93	93	93
26̄ C 2	26	25	24	21	17	13	8	3	3	8	13	17	21	24	25	26
24̄ C 4	24	22	16	7	3	12	19	23	23	19	12	3	7	16	22	24
9 C 6	9	7	3	3	3	9	7	3	3	7	9	3	3	3	7	9
4 C 8	4	3	0	3	4	2	1	4	4	1	2	4	3	0	3	4
10 C 10	10	5	5	10	5	5	10	5	5	10	5	5	10	5	5	10
5 C 12	5	2	4	4	2	5	2	4	4	2	5	2	4	4	2	5
17 C 14	17	2	17	5	16	8	14	11	11	14	8	16	5	17	2	17
8 C 16	8	1	8	2	7	4	6	5	5	6	4	7	2	8	1	8
	96	80	66	90	102	69	42	67	99	92	71	80	116	144	150	152
39 S 2	0	8	16	23	29	34	37	39	39	37	34	29	23	16	8	0
88 S 4	0	36	65	84	88	76	52	18	18	52	76	88	84	65	36	0
10 S 6	0	6	10	10	6	6	6	10	10	6	6	6	10	10	6	0
4 S 8	0	3	4	2	1	3	4	2	2	4	3	1	2	4	3	0
33 S 10	0	29	29	0	29	29	0	29	29	0	29	29	0	29	29	0
9 S 12	0	9	5	5	9	0	9	5	5	9	0	9	5	5	9	0
6 S 14	0	6	1	6	2	5	4	4	4	4	5	2	6	1	6	0
	0	23	40	62	82	73	18	61	103	96	73	74	100	108	73	0
Difference	96	57	26	28	20	4̄	24	6	4̄	4̄	2̄	6	16	36	77	152
Sum	96	103	106	152	184	142	60	128	202	188	144	154	216	252	223	152
x (30ths)	30	29	28	27	26	25	24	23	22	21	20	19	18	17	16	15

The results of all the preliminary tables for the different values of h are collected in a Table IIIc, not shown. This table then serves for the construction of the final tables, of which Table IV shows the portion for $y = 7/60$.   Each set of strips has its amplitude column taken from Table IIIc, and the difference and sum of the cosine and sine parts gives a line of the final totals.

§ 6. *Acknowledgments*

We desire to express our thanks to Mrs. I. Lipson, Mr. B. Boren, and Mr. G. King for help in compiling experimental sets of strips; to Dr. W. Berg for much practical advice; and to Mr. R. W. James for revising this paper and suggesting many valuable improvements.   Also we wish to thank Prof. W. L. Bragg, without whose constant interest and encouragement the method would not have been fully developed.

8. The Structure of Organic Crystals*

PROF. SIR W. H. BRAGG

Summary

For many reasons the structure of crystals of organic substances invites examination by the methods of X-ray analysis; but their molecular complexity would seem to throw great difficulties in the way. It is possible, however, that the difficulties in the case of aromatic compounds may be surmounted by adopting a certain hypothesis, viz., that the benzene or naphthalene ring is an actual structure, having definite size and form, and that it is built as a whole into the organic substances in which it occurs. Reasons can be given why this is *a priori* probable.

The examination of certain organic crystals has been made. The results are in general agreement with the hypothesis, and lead to various deductions of interest.

The structure of organic crystals offers a very inviting field of research by the methods of X-ray analysis. To the organic chemist the relative positions of the atoms in the molecule, as also of the molecules in the crystal, are of fundamental importance; and it is with these relations that the X-rays deal in a manner which is new and unique. Moreover, the multiplicity of crystalline forms—and this is true of both organic and inorganic substances— each so precise and invariable, and so obviously related to the atomic and molecular forces, is a sign that if the forces were better understood it would be possible to account for the forms that are known, and possibly to build others that are unknown. But in order to acquire such a power we must learn the crystalline structure, so that the physical characteristics of the whole may in the end be referred to the characteristics of the individual

*Presidential Address, delivered November 11, 1921. *Proceedings of the Physical Society*, **34**, 33–50 (1921).

atom. Progress has been made with the examination of the structure of some of the simpler inorganic crystals; but organic crystals have been neglected. Their molecular complexity has been somewhat of a deterrent. Yet, if a way could be found of making determinations of structure, in spite of the complexity, it seems likely that they would quickly be fruitful. The substitutions and additions which are so characteristic of organic chemistry take place in such an immense variety of combinations and grades that the slightest knowledge of the underlying mechanism might lead to useful comparisons and rules.

I have made a careful study of a few crystals, principally naphthalene and some of the naphthalene derivatives, in order to discover, if possible, some way of handling the complex molecules. The numerical results will be set out later, and may, I think, be taken as sufficiently accurate to make foundations for a theory.

I shall endeavour to show that the results can be explained, so far as can be seen at present, by supposing the benzene ring or naphthalene double ring to have definite form and size, preserved with little or perhaps no alteration from crystal to crystal, and that there are good *a priori* reasons for the supposition. If this principle be accepted the problem is simplified at once. Naphthalene itself is then to be regarded as a structure in which there is but one element, the naphthalene double ring, and no longer as an aggregate of 10 carbon atoms and eight hydrogen atoms of unknown mutual arrangement. A more complex molecule such as either of the naphthols is not to be regarded as an addition of one oxygen atom to these 18, an idea on which nothing can be built, but as a naphthalene double ring of the same size and form as before, except that one particular hydrogen has been replaced by a hydroxyl group. It is then possible to think what changes in the disposition of the molecules might be caused by such a substitution and to compare conceivable solutions with observations on the dimensions of the new crystal. Such a method of procedure is obviously in good agreement with the ideas of organic chemistry.

It is convenient to distinguish the facts regarding crystalline structure which can be obtained by the goniometer and various other means, from the new facts which can be obtained by the use of X-rays. The former are recorded in crystallographic tables such as are given by von Groth in his "Chemische Krystallographie". Naphthalene may be taken as an example. In the fifth volume of von Groth's work, on p. 363, a description

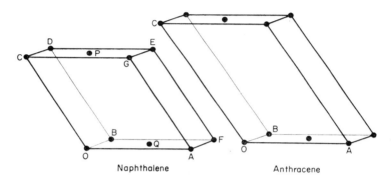

FIG. 1. Unit cells of naphthalene and anthracene drawn to the same scale.

		$OA=a$	$OB=b$	$OC=c$
Naphthalene	....	8·34	6·05	8·69
Anthracene	....	8·7	6·1	11·6

Naphthalene $\alpha=BOC=90°$, $\beta=COA=122°\ 49'$, $\gamma=AOB=90°$
Anthracene $\alpha=BOC=90°$, $\beta=COA=124°\ 24'$, $\gamma=AOB=90°$

of naphthalene is given from which the following data are taken:—

MONOCLINIC PRISMATIC

$a:b:c = 1·3777:1:1·4364;$

$\beta = 122°\ 49'$

The monoclinic prismatic class has the highest symmetry of which the monoclinic system is capable, having a digonal axis and a plane of symmetry perpendicular to it. The figures give the angular, but not the linear, dimensions of the unit cell of the structure (Fig. 1); the unit cell being the smallest volume, of which, by continual repetition without any change in contents or disposition, the whole crystal can be formed. In any crystal it is possible to choose the unit cell in many ways, but they must all be capable of derivation from one another and must all have the same volume. The angular dimensions are to be considered as including the angular relations to one another of any pair of planes in the crystal, not merely of the planes bounding the cell.

It is also stated that the (001) face is the cleavage plane; and that, in addition, the faces (110), (20$\bar{1}$), (11$\bar{1}$) are found as bounding planes of the crystal. The angles between various pairs of these faces are also given as observed. The specific gravity is stated to be 1·152. Other information is given by von Groth concerning the optical properties of the crystal; also concerning the methods that have been adopted in growing the crystals from various solutions and the consequent effect on the development of different faces. These facts do not concern us for the moment, but they must be taken into account eventually.

The examination by X-ray analysis gives us the spacings between the planes and, therefore, the linear as well as the angular dimensions of the unit cell. The specific gravity being known, and the actual weight of the molecule, it is possible to find how many molecules are contained in each cell; generally, two or four. In the case of naphthalene, it is found that, assuming the angular dimensions to be correctly given by the crystallographers the linear dimensions are:—

$$a = 8\cdot34, b = 6\cdot05, c = 8\cdot69.$$

These figures are obtained in the following way:—

The actual length of the b axis being represented by b, the mass contained in the cell is

$$b^3 \times 1\cdot3777 \times 1\cdot4364 \times \sin 122° 49' \times 1\cdot152 \text{ A.U.}$$

(It is convenient in this work to extend the Angstrom system of units so that an A.U. of area is 10^{-16} cm.^{2}, of volume 10^{-24} cm.^{3}, and of mass 10^{-24} gr.)

The mass of the hydrogen atom being 1·662 A.U., the mass of the molecule $C_{10}H_8$ is $128 \times 1·662 = 213$ A.U. Now, from the full results of the X-ray measurements which will be given presently, it is perfectly clear that there are two molecules in each unit cell. Hence, the value b is readily calculated, and the values of a and c also.

Besides these determinations of length, the X-ray method gives also the angle between any pair of planes, whether they form faces or not, provided that a measurable reflection can be found. Also, the relative intensities of the reflection by different faces, as well as the relative intensities of the spectra of different orders given by any one set of planes, yield information as to the distribution of the scattering centres and of the atoms which contain them.

There are two distinct methods of using the X-rays. In the one, which has been used from the beginning, a single crystal is employed. It need not usually weigh more than 2 or 3 milligrammes, and, in fact, it is more convenient that it should be small, since the pencil of reflected rays is then conveniently limited by the size of the crystal without the need for slits, which are still used, however, but as stops only. This method is at present the more precise of the two. Also it permits the measurement of the angles between reflected planes, a knowledge of which is often very useful for identifying the planes.

The second method, first used for crystal analysis independently by Debye, and by Hull, can be used when the crystal is in powder, and can, therefore, be employed when no single crystal can be obtained of sufficient size to measure by the first method. All the spectra of the different planes are thrown together on the same diagram or photograph, and must be disentangled. This is not so difficult as it might seem, because there are not more than one or two lines in each spectrum, and there is generally independent evidence as to the angular dimensions of the crystal. The spectra of copper, and of some organic substances, shown in Figs. 2, 3, 4 and 5, were obtained in this way. The

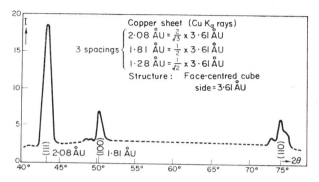

FIG. 2. Copper spectrum, obtained by powder–ionisation method.

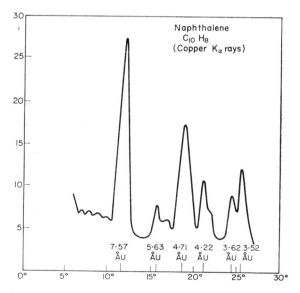

FIG. 3. Spectrum of naphthalene–powder method. The results agree well with those obtained by the single crystal method, but are slightly larger.

G*

X-ray bulb employed was of the form described by Müller.[1]
The way of using the powder in conjunction with an ionisation
spectrometer has been described in the Proceedings of this
Society.[2]   Müller's bulb is very well suited to this method.
The spectrometer slit can be brought very close to the radiator,
so that the slit becomes the source of a strong beam of rays
which is sufficiently divergent to cover a large surface of the plate

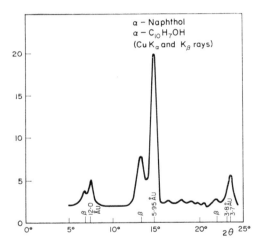

FIG. 4. Spectrum of α-naphthol–powder method.

on which the powder is spread. By a known focussing action
(X-rays and Crystal Structure, W. H. and W. L. Bragg, p. 31)
the rays are converged upon the slit of the ionisation chamber.
The anticathode used in these experiments was made of copper;
the long wave lengths of the K series of copper give suitable
angles of deflection even for the wide spacings that are found in
organic crystals. The curve in Fig. 2 shows the spectrum of copper

---

[1]*Phil. Mag.*, p. 419, Sept. (1921).
[2]*Proc. Phys. Soc.* **33**, [4], 222, June 1921.

obtained by the use of copper-rays, and gives at once all the necessary information from which the structure of the copper crystal is derived. The piece of copper employed was of ordinary sheet, not prepared in any way.   The spectra of the organic substances show how very diversified they are, and illustrate the power of a method of analysis which promises to be of great use,

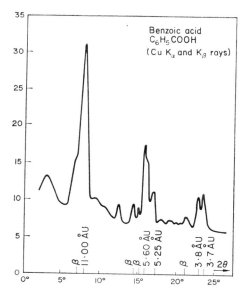

FIG. 5. Spectrum of benzoic acid powder method.

since every crystal has its own characteristic spectrum. The bulb was driven by a transformer capable of developing half a kilowatt; but not working to half of its full capacity.   Each determination of a point on the curve took about 10 to 30 seconds.

We may now consider what reasons can be put forward for assuming the concrete existence of the benzene and naphthalene rings.   If we examine the structure of the diamond we find that the atoms of carbon are tied together so that each is at the centre of gravity of four others. The distance from centre to centre is 1·54 A.U. As I have already pointed out[3] the rigidity of the diamond and the open character of its structure, imply that great force is required to alter the orientation of any coupling with respect to the other three belonging to the same atom.

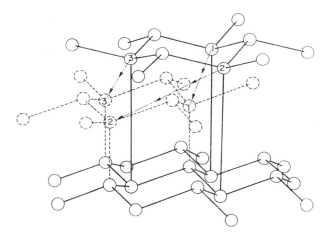

FIG. 6. The fine lines of the diagram show the structure of graphite. By moving the top layer to the position shown by the broken lines the diamond structure is obtained.

Were it otherwise, all atoms would seek to be surrounded by as many neighbours as possible; the substance would be close packed, and its density would be more than double what it is. The structure of the diamond may also be looked on as consisting of a series of puckered layers parallel to a given tetrahedral plane (*see* Fig. 6). A sharp blow may cleave the diamond, along one of these layers. If we take a model showing two layers as in graphite, lay hold of the upper layer, and move it to the new

[3] *Proc. Phys. Soc.* **33**, [5], Aug. 15, 1921.

position shown in the figure, the structure is now that of diamond. I am following Hull's determination;[4] Debye and Scherrer[5] would flatten out the puckers in the planes of graphite. The point is perhaps not important for our present purpose, but it is necessary, for descriptive purposes at least, to choose one form. According to Hull's measurements, the shortest distance between each pair of atoms lying in the same layer is shortened from 1·54 A.U.—the diamond spacing—to 1·50 A.U. The distance between two successive layers has been increased by 1·35 A.U. A carbon atom in one layer is now at equal distances from its three nearest neighbours in the next layer, the distance being 3·25 A.U.

The bonds between one layer and the next are now greatly weakened; the substance cleaves readily in thin flakes. One layer slides with great ease over the other, though the bonds between the atoms in any layer are at least as strong as before. When all the bonds were of the strong kind, the substance, as diamond, was the hardest thing known. When the one set of bonds has been weakened, the substance, as graphite, is used as a lubricant. Probably its efficiency as such depends both on the weakness of the one set of bonds and the strength of the other. Yet these new bonds are perfectly definite, and the distance between two layers and, therefore, the 3·25 distance between atomic centres is a perfectly constant and determinate quantity. It should be of great interest to compare the physical constants of graphite along and perpendicular to the axis, since in the two cases the two kinds of bonds are separately involved. Some of the comparisons would be difficult, but the thermal expansions can probably be compared by X-ray methods.

If the strong bonds between the atoms in a layer remain, and are even drawn a little tighter when the graphite form replaces that of diamond, it seems very reasonable to suppose that the single ring or multiple rings which are so clearly to be distinguished in the network may be separated out as such without

---

[4]*Physical Review*, **10**, 692, Dec. 1917.
[5]*Phys. Zeit.* **13**, 297 (1917).

loosening the bonds between their component atoms. In fact, these latter bonds might be expected to tighten even a little more. Let us assume that a single ring is a benzene ring, a double ring a naphthalene ring, and so on. Taking the spacings as given by Hull for graphite, the dimensions of the benzene and naphthalene frameworks are as shown in Fig. 7. The figure is constructed to show the arrangement of atoms in the naphthalene crystal, but it will also serve to illustrate the point under discussion. The carbon atoms A to F form a benzene ring, those from A to J the double ring of naphthalene.

The atom centres A, and G, are 0·71 A.U. above, and the centres D and J the same distance below the plane of the diagram which is supposed to contain all the remaining centres. It should be observed that circles are used to represent the atoms as a convenient method of designation, not as implying that the radius (1·50) may *always* be used in calculating the distance between the centre of any one atom, and the centre of any other atom.

We may now go on to consider individual crystals; and we take naphthalene first. It might have seemed more natural to attack the benzene crystal before the naphthalene; but the latter was chosen because it is a very well-shaped crystal, and is solid at ordinary temperature. Benzene can only be examined under special temperature conditions and then only, with convenience, as a mass of small crystals. The study of benzene and some of its derivatives has been begun, but the greater attention has been given to the naphthalene crystals, and I will describe now the results that have been obtained in their case.

In the second column of Table I are given the angles between certain pairs of planes as calculated from the data already given, viz.:—

$$a = 8\cdot34, \ b = 6.05, \ c = 8.69, \ \beta = 122° \ 49',$$

and in the third the corresponding values as observed.

These figures are quoted in order to show that the X-ray method gives angular measurements with sufficient accuracy.

Indeed the agreement between the values in the two columns is as good as is usually reported by different observers who have used the goniometer (*see* the descriptions recorded by von Groth); and the X-ray method could really be relied on to a

TABLE I

Angles between	Calculated	Found
100:001	57°11′	57°16′
110:11$\bar{1}$	36°32′	36°34′
20$\bar{1}$:001	85°45′	85°48′
021:001	67°47′	67°35′
20$\bar{1}$:10$\bar{1}$	30°21′	30°23′
210:21$\bar{1}$	24°12′	24°24′

TABLE II

Plane	Calculated spacing	Observed spacing	Nature of reflection
100	7·00	3·46	Strong: indication of spacing 6·92
010	6·05	2·95	Very weak
001 (Cleavage)	7·30	7·30	Very strong: also higher orders
110 (Natural)	4·59	4·55	Strong
11$\bar{1}$ (Natural)	4·70	4·63	Moderate
20$\bar{1}$ (Natural)	4·17	4·12	Strong
021	2·79	2·76	Very weak
10$\bar{1}$	7·51	3·71	Very weak
210	3·04	2·99	Strong
21$\bar{1}$	3·44	3·39	Strong

minute of arc if the necessary care and time were justified. In this case no special pains were taken.

In the next table are set out the results of the linear measurements. They are compared with calculations based on the assignment of *one* molecule to each cell, which is equivalent to supposing each corner of the cell to represent a molecule. The object of making this provisional assumption is to show how the position of the second molecule can be found by comparing the observed with the calculated results.

No reflection obtained from following planes:—

011, 012, 10$\bar{2}$, 101, 111, 22$\bar{1}$, 112, 11$\bar{2}$, 221, 21$\bar{2}$, 211, 212.

The table shows that the 100 and 010 spacings are only half what they should be if there were molecules at cell corners only. (N.B.—Only one-eighth of a corner molecule is within the cell, and the whole eight count for one whole molecule within the cell). But the 001 spacing is right. We conclude that there is a molecule at each of the points P and Q (*see* Fig. 1), each contributing half a molecule to the cell: and that these are in all respects similar to the corner molecules. Molecules placed at P and Q interleave the planes 100, 010, and also 10$\bar{1}$ by other planes of equal density which halve the corresponding spacings. The planes 110, 20$\bar{1}$, 021, are unaffected because they already contain P and Q.

It should be mentioned, however, that the 100 plane seemed to give a small spectrum at half the angle which gave the principal reflection; this would indicate that the second molecule was not quite similar, in orientation or some other particular, to the first, or was not exactly at P and Q. This is also suggested by the fact that 210 and 21$\bar{1}$ give the calculated spacings, whereas half values might be expected.

Although so much information is given in these tables, some of which we have used and some of which we cannot fully use for lack of knowledge, yet it would be hopeless to try to arrange the eighteen atoms of naphthalene on the basis of what has been learnt, without some helpful hypothesis. But we now take the naphthalene double ring as described. Its dimensions are such that it seems quite possible to fit two of them into the cell, if we had some indication as to their orientation thereto.

As to this we get a strong hint from a comparison of naphthalene with anthracene ($C_{14}H_{10}$), whose construction shows three rings in a line, as against the two of naphthalene. The crystallographic data of the latter are:—

$$a:b:c = 1 \cdot 4220:1:1 \cdot 8781, \ \beta = 124° \ 24'.$$

Specific gravity = 1·15.

The crystals themselves are very small flakes, and it was not possible to find one which could be conveniently treated by the single crystal method. However, by pressing a number of them together against a flat disc, so that all the 001 planes were parallel thereto, however oriented they might be otherwise, it was easy to get a sufficiently accurate determination of the 001 spacing, and therefore the linear dimensions of the unit cell. It appeared that there were two molecules to the cell, as for naphthalene. The dimensions were:—

$$a = 8 \cdot 7, \ b = 6 \cdot 1, \ c = 11 \cdot 6.$$

If these dimensions are compared with those of naphthalene (*see* Fig. 1), it will be seen that while a, b, and β remain nearly the same, the c axis has lengthened considerably, the difference amounting to 2·9 A.U. nearly. Now the extra ring, if of the benzene dimensions, should be responsible for an additon of 2·5 A.U. nearly to the molecule.

It is reasonable to conclude that the molecules in both crystals lie end to end along the c axis, and that the structures are similar.

The over-all lengths of the two molecules, without allowance for the hydrogen atoms at their ends, that is to say, in the β-positions, are 6·41 and 8·86 respectively. There is, therefore, a vacant space between the ends of two molecules of rather more than 2 A.U., into which two hydrogen atoms have to be fitted. This agrees very well with what might be expected; only it must be remembered that we have no definite indication from studies of crystal structure as to the actual distance between the centres of a carbon and a hydrogen when united by a valency bond, nor between two hydrogen atoms not so united.

We have still to decide in what plane, passing through the c axis, the molecule is to be placed, and we have less clear indications with respect to this point than those that have guided us hitherto. On making up a model, however, it is seen that it is much more likely that the plane of the molecule lies nearer to the *ac* plane than the *bc* plane. The molecules lock together much better if that is so. Moreover, if the molecules lie in the *bc* plane, they would be close neighbours in that plane, and at the same

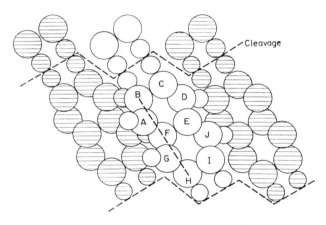

Fig. 7. Showing mutual relations of three naphthalene molecules and parts of others.

The unshaded circles between the two cleavage planes represent a molecule as at Q (Fig. 1). The shaded represent molecules B and F in the same figure. The small circles represent hydrogen atoms, but their size is uncertain.

Diameter of carbon atom $= 1 \cdot 50$. $BH = 4 \cdot 92$.   Projection of AD on the plane of the diagram $= 2 \cdot 50$.

time there would be wide gaps between consecutive planes. The plane 100 should, therefore, be prominent, most probably a natural face, perhaps even a cleavage plane: whereas it is neither of these things. But if the molecules lie in the 010 plane the form of the crystal seems much easier to understand, as we shall see later. The β-hydrogens of each molecule lie up against the corresponding hydrogens of the next and the 001 plane passes through them all. It would appear to be the weakest junction in the crystal, and therefore the 001 plane is the cleavage plane.

It must be observed that in the junctions between molecule and molecule there are forces far weaker than the valency forces, which latter unite the atoms of the same molecule. It is the former which bind the molecules into the crystal, nevertheless.

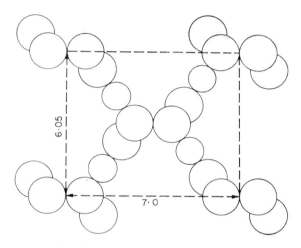

FIG. 7a. Section of naphthalene cell perpendicular to the axis of c, showing α-hydrogens connecting the molecules side to side.

When the model is put together in the way now indicated it is found that all the α-hydrogens, those that are attached to the sides of the molecule, lie up against carbon atoms of the next molecule, and that there is an appropriate space waiting for each, of magnitude about 1 A.U., the actual value depending on the orientation of the molecule. The forces exerted at these junctions, though far weaker than valency forces, are stronger than the forces between two β-hydrogens; and therefore if the crystal is ruptured it is the latter which give way first. The forces exerted by the α-hydrogens do so across the 110 and 1$\bar{1}$0 planes, and it is not surprising to find that the latter make natural faces of the crystal and give a strong reflection. The structure now found is a very empty one: it is like lace-work in space. That must be expected, since the specific gravity is so low. The structure is shown in Figs. 7 and 7a.

We may now attempt to find the structure of a naphthalene molecule in which some complexity has been introduced by a substitution. We take acenaphthene. Dr. J. B. Cohen was kind enough to give me some good crystals weighing each a few milligrams.

TABLE III

Plane	Calculated spacing	Observed spacing	Remarks
100 (Natural)	4·16	4·16	Moderate strength
010 (Cleavage)	7·07	7·1*	—
001	3·63	3·63	Moderate strength
110 (Natural)	3·57	7·18	Moderate strength
101	2·74	2·74†	Very weak
011	3·23	3·19	Very weak
111 (Natural)	5·10	5·1	Weak

*Not observed directly, but inferred from other observations.
†There is an indication that alternate planes differ somewhat.

In this case the molecule has been made lop-sided by the substitution of a group of two carbon and four hydrogen atoms for the two hydrogens on one side. It is a striking fact that the crystal has now gained in symmetry; the unit cell is rectangular. The crystallographic data are:—

Orthorhombic (with the fullest symmetry of that class)—

$$a:b:c = 0.5903:1:0.5161.$$

It appears that there are four molecules to the cell, and that

$$a = 8.32, \quad b = 14.15, \quad c = 7.26 \ (see \ \text{Fig. 8.})$$

The doubling of the number of molecules, as compared with naphthalene, suggests that symmetry has been obtained by an arrangement in which two of the unsymmetric molecules are the images of the other two across one of the principal planes, *ab*, *ac*, or *bc*.

As the symmetry is so high, it is natural to test first the hypothesis that there is a molecule at each corner of the unit cell, and

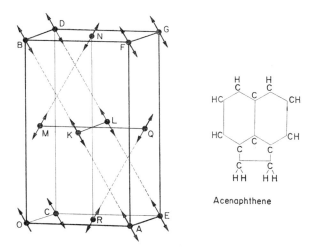

Acenaphthene

FIG. 8. Unit cell of acenaphthene, containing four molecules.
$OA=8\cdot32$, $OB=14\cdot15$, $OC=7\cdot26$.

one in the middle of each face, which arrangement will account for four molecules. The calculated spacings in Table III are in accordance with this supposition.

There is complete agreement, except for the 110 plane. Assuming that there will be an explanation of this as we proceed further, we now look where a molecule of length about 8·69 (as in naphthalene) may be placed in position, and observe that $KB = KF = \sqrt{(4\cdot16^2 + 7\cdot07^2)} = 8\cdot23$, and these two are images of each other in the 100 plane. This is not quite long enough.

Let us, however, suppose provisionally that the molecules at the corners of the cell and at K and L lie parallel to AB, and with their planes coincident with (001). Let those at $MNQR$, which between them contribute to the cell the other two molecules that belong to it, lie in the parallel plane, but slope the other way, parallel to MN or QR. If this is so, we see why the 110 plane $ABDE$ in the figure seems to have a double spacing; it is because the two sets of molecules—viz., such as A, B, D, E of the one set, and M, N, Q, R of the other, are very differently disposed towards this plane. One set lies parallel to it, the other not. Consequently the spacing is twice what it would have been had the molecules been all alike with respect to this plane, and is equal to the perpendicular from F or G upon the plane $ABDE$. This is true of the 110 plane alone, out of those set out in the table above.

Once more the cleavage plane (e.g., $BDGF$) passes through the junctions of β-hydrogens. On constructing a model it is found that the projecting addition on the side of a molecule seems to engage with a neighbouring molecule on that side of the latter on which its own addition does not lie. The crystal is thus laced together; and in lines which for alternate layers parallel to the cleavage plane (010) are parallel to OE and AC respectively.

The two points left doubtful are the small difference between the length of the molecule in this crystal and in naphthalene, and the fact noted with respect to the plane 101 (see the table).

We now take α-naphthol. Here one of the hydrogens at the side of the molecule has been replaced by OH. It was difficult to obtain without special preparation a crystal large enough for the single crystal method; but on sorting over a considerable amount of material a few flakes were found, each two or three millimetres across and a fraction of a millimetre thick. With one of these some very good observations were made, as recorded in Table IV. The crystallographic data are:—

Monoclinic prismatic.

$$a:b:c = 2 \cdot 7483 : 1 : 2 \cdot 7715, \ \beta = 117° \ 10'.$$
$$\text{Specific gravity} = 1 \cdot 224.$$

It is natural to suppose that the similarity of these figures to naphthalene, when the *a* and *c* have been halved, means that the naphthalene cell has been increased four times in order to allow all the four α positions to be represented in due order. But it turns out that the crystal is of a totally different structure. There are four molecules in the cell (*see* below), as defined in accordance with the above data. If the naphthalene cell had been repeated four times, there would have been eight molecules in the larger cell.

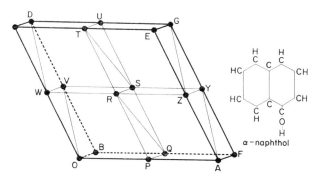

FIG. 9. Unit cell of α-naphthol containing four molecules.
$OA=13·1$, $OB=4·9$, $OC=13·4$, $\beta=117° 10'$.

The linear dimensions are:—

$$a = 13·1, \ b = 4·9, \ c = 13·4 \ (see \ \text{Fig. 9})$$

and it is clear that the molecules are placed as in the figure, one at each point indicated by a letter. This is, perhaps, seen most readily if the calculated figures of the following table are reckoned as if there were one molecule at each corner of the large cell only—viz., *OAFBCEGD*. We then have the comparison as shown in Table IV.

Also angles between   210, 21$\bar{2}$ 36° 24′ (calc.)   36° 16′ (obs.)

and between   10$\bar{1}$, 100   57° 48′   ,,   58° 5′   ,,

In the cases where a bracket occurs in this table, it is meant that a reflection shows the existence of a certain spacing between planes, while a very strong reflection at twice this glancing angle shows that this spacing is halved by other planes differing considerably from the first. See, for example, Fig. 4, where a small peak shows the 001 spacing 12·00, and a much larger peak at twice the angle corresponds to half the spacing. This agrees with a supposition that there are molecules at every point indicated by a letter, but those at *VSYZRW* are very different in some way from the others.

TABLE IV

Plane	Calculated spacing	Observed spacing	Remarks
Cleavage 001	11·92	12·00*	Strong
002	5·96	5·95*	Very strong
210	3·75	3·76	Weak
21$\bar{2}$	3·69	3·68	Rather weak
100	11·82	—	—
200	5·91	5·84	—
10$\bar{1}$	11·38	—	—
20$\bar{2}$	5·69	5·62	†

*Powder method only.
†The symbol 002 is used to denote the second order reflection of 001. When it is abnormally strong, it follows that the 001 planes are interleaved.

The reflection of 100 in the powder photograph would be indistinguishable from 001, and perhaps also the 10$\bar{1}$. The powder method does not give very clear indications, therefore, of the existence of the spacings 11·82 and 11·38, because all that is observed may be due to the 11·92, which must be strong, as 001 is the cleavage plane. The single crystal method gave the proper reflection of the 200 and 20$\bar{2}$ planes, as shown in the table; but the reflections belonging to the double spacings 100 and 10$\bar{1}$ were not looked for, and probably would have been very hard to find with so small a crystal. However, the combined data are amply sufficient for the conclusion that the four cells in the figure really imply four different orientations of the molecule, and that if we place molecules exactly at the corners of the large fourfold cell the other molecules are very nearly at all the other corners.

We have this further indication that, though the half of the c axis, $OW = 6.7$, is too short to take in the naphthalene molecule, yet the distance OV is equal to 8·3, exactly the distance into which the molecule seemed to fit in acenaphthene. Once again, then, it would seem that these lopsided molecules lie criss-cross, as represented by the lines parallel to OV in the figure; but they now lie edgeways on top of one another, not flatways, as in the other crystal. It is the "a" axis that now runs along the line of crossings, not the "c" axis. The cleavage plane again passes through the β-junctions.

When a model is made, it is found that the hydroxyl groups fit into their places very naturally, and that if they link together the tops of the molecules in any one layer parallel to the 001 plane, they link in the same way the bottom ends in the next layer. This brings pairs of hydroxyl groups rather close together, as one might expect; and it looks as if the attraction existed specially between the two oxygens. The attraction is exerted across every alternate 001 plane. The other 001 planes are the cleavage planes. All the other linkings appear to be done by α-hydrogens.

The crystallographic data of β-naphthol are:—

$$a:b:c = 1.3662:1:2.0300; \quad \beta = 119° 48'$$

$$\text{Specific gravity} = 1.217.$$

This has been examined by the powder method only, as a single crystal was not readily obtainable. There is a very strong reflection for the spacing 7·90. This is sufficient to fix the linear dimensions of the cell. If we put

$$a = 5.85, \ b = 4.28, \ c = 8.7 \ (see \ \text{Fig. 10})$$

the weight of material in the cell is equal to that of one molecule, and the 001 spacing is 7·76, in close agreement with what is found.

Comparing these figures with those for the smaller cell of α-naphthol such as $OBQPWVSR$ (Fig. 9) which are 6·55, 4·9 and 6·7 respectively, it is clear that the removal of the OH group from the side to the end of the molecule has caused the cell to shrink sideways in both a and b directions and to grow along the c axis; this confirms the hypothesis that the molecules in these crystals lie mainly along the c axis. The cleavage plane still cuts across the β-junctions.

The question naturally arises as to whether this is really the unit cell. Ought it not to be multiplied by four, as for α-naphthol, because there are four β-positions, as well as four α-positions? The evidence in favour of the extension seems very strong. In the first place, it would be the only case, if the present crystallographic data were correct, in which a cell contained one molecule only. Moreover, if all the molecules were oriented alike, their dissymmetry would lower the symmetry of the crystal. The symmetry which is assigned to it can only be obtained by making the unit cell large enough to contain four molecules so that all the β-positions may be represented. Also the hydroxyl groups will surely tend to draw together and cause pairs of molecules to point opposite ways. For these reasons alone it seems likely that the structure is really the same as that of α-naphthol. The length OV which was 8·3 for α naphthol becomes 9·7, a growth of 1·4 which seems consistent with the value of the diameter of the oxygen atom, viz., 1·30.[6]

No other naphthalene derivatives have yet been examined by the X-ray methods, but there are, of course, large numbers that may be expected to give useful results. It is worth remarking that the unit cell of α-naphthylamine has almost exactly the same dimensions as that of acenaphthene, if it be supposed to

---

[6] W. L. Bragg, *Phil. Mag.*, p. 169, Aug. 1920.

contain four molecules. Both are rectangular; the acenaphthene dimensions are:—

$$a = 8.32, b = 14.15, c = 7.26,$$

those of α-naphthylamine are:—

$$a = 8.62, b = 14.08, c = 7.04,$$

provided that the a and b axes of the crystallographers are interchanged in the latter crystal. Their habits also correspond closely, and although the latter crystal has no clear cleavage

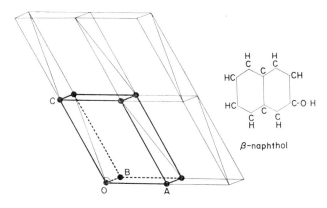

β-naphthol

FIG. 10. Unit cell of β Naphthol, in thick line, containing one molecule.

$$OA = 5.85, \quad OB = 4.28, \quad OC = 8.7, \quad \beta = 119° \, 48'.$$

plane it has a very prominent (010) face which may represent the 010 cleavage of the former. Assuming that the arrangement is therefore the same in the two crystals, the length of the molecule in the new case is $\sqrt{(4.31^2 + 7.04^2)} = 8.25$; as against 8.23 for acenaphthene and 8.31 for α-naphthol. In all these three cases the lopsided lines of molecules are laid criss-cross upon one another, whereas in naphthalene the lines are parallel to each other. The somewhat greater length of the molecule in napthalene, viz., 8.7, may be connected with the difference in arrangement.

I have made some measurements of benzene compounds, but the analysis of the results is incomplete; and until benzene itself is examined data are wanting that must be important. It is worth mentioning, however, that benzoic acid has a very wide spacing between the (001) planes. The dimensions of the unit cell containing four molecules, are:—

$$a = 5\cdot44,\ b = 5\cdot18,\ c = 21\cdot8,\ \beta = 97°\ 05'$$

Experiment by the powder method shows that there is a very strong 10·9 spacing implying that the sheets of molecules lie at this distance apart parallel to the 001 plane. If the crystallographic data are correct these sheets differ alternately. Even a 10·9 spacing is very wide, and this suggests that extended *COOH* groups are required to bridge across the intervening space. Such a bridge may be made, of course, by a *CO* extension from one ring joining on to an *OH* extension from another ring. These bridges part very easily, it would appear; the crystal flakes at the least touch.

The work that I have now described may conveniently be divided into two parts. The first is experimental, and shows, I think, that a large mass of new and valuable information respecting the linear dimensions of organic crystals is to be gained by the methods of X-ray analysis. In the second, I have ventured to suggest and apply a certain principle, viz., that the benzene and naphthalene rings as well as other ring combinations have actual form and dimensions which are nearly, if not quite, the same when they are built into different compounds. Such a principle requires much more illustration before it can be finally established; for the present, however, it shows promise of providing an entry into a very wide field which otherwise seems difficult to break into.

Moreover, the idea fits in extremely well with the work of Langmuir on surface films, and with such results as those described by Adam,[7] and with the ideas and experimental

---

[7] *Proc. Roy. Soc.* **99**, July (1921).

results of several other workers.[8] We see clearly that the forces that bind atoms together are of more than one kind. The very strong valency bonds, whether explained as due to electron sharing, or in any other way, are exemplified by all the linkings in diamond and by the linkings in the planes of the graphite flakes. But, besides these are other bonds of much weaker character, such as those extending between an atom in one graphite sheet and its three nearest neighbours in the next. Such bonds as these unite the molecules of the organic compound so as to form the crystal. They are of varying degrees of strength; the cleavage plane shows where they are weakest. They are definitely associated with special points on the molecule as we see from the facts of crystallisation. When a crystal forms in a liquid, or by sublimation, the molecule that attaches itself correctly, and with proper orientation to others already in position, is the one that stays there and resists the tendencies of other drifting and thermally-agitated molecules to remove it. It is fixed by the attachment of certain points on its own structure to certain points on the structure of the other molecules. The beautiful exactness of crystal structure is evidence of the precision with which this adjustment is made; and at the same time of the definite molecular form without which precision would be impossible.

The molecules of naphthalene arrange themselves side by side, the α-hydrogens of each molecule seeking to attach themselves to the carbon atoms of its neighbours; valency bonds do not enter into this. Whether there is some special tendency to make an attachment with a carbon atom in the centre of the molecule as seems to be the case, or whether the general slant of the molecules is due to some other cause, is a matter for further inquiry. But it is the side-to-side combination that is preferred in comparison with the end-to-end; the former gives up more energy than the latter. Consequently the crystal grows out quickly into thin sheets. We may further imagine that sometimes, and more often in certain crystals than in others, as well as more often with respect

---

[8] See a list given by Adam, loc. cit., p. 351.

to one plane in a given crystal than to other planes, the geometry may permit a molecule to attach its active points to those of other molecules, already settled, in more than one way which does not carry on the regularity of previous attachments. The crystal starts a new growth, and twinning results.

The surface films studied by Langmuir and by Adam and others were not formed of aromatic compounds such as the few that have been investigated as crystals. But the strength of the side-to-side attachment of their molecules as compared with the end-on attachments seems to be exactly the same in effect and to lead to similar results. In a mass of oleic acid the long molecules are lying in all ways, and their heat motions do not give them a chance of attaching themselves at sufficient points at the same instant. But when the hydroxyl ends of some of them root themselves in the water, these molecules are held quiet for a moment and the others quickly slip into their places and link on side to side. So the film spreads out quickly over the water. It is a kind of catalytic action; the water does not itself draw the film flat, but helps the molecules of the film to arrange themselves. In a sense, the film has negative surface energy.

The arrangement of molecules in crystals or in surface films or interfaces between liquids and solids cannot be fully explained as due to forces which are merely functions of the distances between their centres. Confining ourselves to the cases where there is no obvious separation of electron charges, as there is none in the crystals described above, it is clear that we must think of the molecules as bodies of very definite form. These attachments to one another are made at definite points and the forces there exerted may have very short ranges. The molecules are locked into crystalline structure, when attachments are made at sufficient points, and the whole has the stability of an engineering structure. It may well be that in a liquid there are always some completed attachments, but insufficient in number to give rigidity to the whole. In a gas there are no attachments at all.

I have to thank Mr. Shearer for his assistance in examining crystals by the powder method, which method he has greatly improved; and Mr. J. Reid, for the labour and skill which he has devoted to making models.

Part of the apparatus employed was obtained through a grant from the Department of Scientific and Industrial Research.

I should also like to thank the General Electric Company of America for their great kindness in presenting me with Coolidge X-ray bulbs of special design.

Dr. C. Chree, in moving a vote of thanks, said that the President's researches had opened up fascinating vistas of investigation, which should lead to momentous discoveries.

Sir R. Robertson, expressing the hope that these accurate measurements and brilliant speculations would in time be applied to the compounds of nitrogen, seconded the vote of thanks, which was carried by acclamation.

9. An X-ray Study of the Phthalocyanines, Part II Quantitative Structure Determination of the Metal-free Compound*

J. Monteath Robertson

Preliminary data on the crystal structure of some of the phthalocyanines have already been given (J., 1935, 615). These compounds (see Linstead, J., 1934, 1016) form an isomorphous series of remarkably stable monoclinic crystals. The present paper gives the results of a quantitative determination of the structure of free phthalocyanine, $C_{32}H_{18}N_8$, space group C_{2h}^5 ($P2_1/a$), $a = 19.85$, $b = 4.72$, $c = 14.8$ Å, $\beta = 122.25°$, which has two centrosymmetrical molecules per unit cell of volume 1173 Å³; d (found) $= 1.44$; (calc.) $= 1.445$, $M = 514$, $F(000) = 532$.

It is well known that the direct application of the X-ray method of analysis is severely limited in the general case by the ambiguity of the phase constant. It is possible to make absolute measurements of the intensities of the X-ray crystal reflections, and from these values to calculate the magnitudes of the structure amplitude factors (F). But the complete expression of F, as it occurs in the Fourier series representing the electron density at any point in the crystal lattice, is a complex quantity characterised by an amplitude and a phase constant. As the X-ray crystal reflections are recorded separately, at different times and with the crystal in different positions, all information regarding the relative values

*Journal of the Chemical Society, pp. 1195–1209 (1936).

of the phase constants is necessarily lost in making the experiment with a single crystal. However, it is usually possible to proceed with the analysis by methods of trial, making use of previous knowledge of the scattering power of atoms and the probable distances between them. The application of the Fourier synthesis then leads to a refinement of the preliminary results, and by a method of successive approximation is equivalent in the end to a direct determination of the interatomic distances. The analyses already carried out in this way for organic structures (*Proc. Roy. Soc.*, 1933, A, **140**, 79; 1935, A, **150**, 348; etc.) are thus to some extent unsatisfactory, not because they are lacking in truth, but because of the rather indirect manner by which the results have been obtained.

It is remarkable that the complex phthalocyanine molecule, governed by 60 independent parameters for the carbon and nitrogen atoms alone, should be the first organic structure to yield to an absolutely direct X-ray analysis which does not even involve any assumptions regarding the existence of discrete atoms in the molecule. We do, however, require some knowledge of the properties of the metal atom in the phthalocyanine derivatives; *e.g.*, we require to know its atomic number and its atomic weight approximately, and we have to assume that most of its electrons are concentrated in a space which is quite small compared to the size of the whole molecule. One other assumption, which follows from the isomorphism of these compounds, is necessary. If the dimensions of the unit cells of two related members of the series are very closely similar, we assume a corresponding similarity in the structures contained within these cells, not necessarily in detail, but in broad outline, especially in the overall dimensions of the related molecules. The remainder of the analysis is perfectly direct, because both the amplitudes and the phase constants of the component terms in the Fourier series for the electron density can be determined from a double series of absolute intensity measurements from the compound with and without a metal atom present. This applies generally to the ($h0l$) zone, which gives by far the most important projection

H

of the structure, and with certain exceptions, noted in a later section, to the other zones.

The structures possess centres of symmetry. The phase constants are thus limited to 0 or π; *i.e.*, either a peak or a trough of each component sinusoidal distribution of density in the structure must coincide with the centre of symmetry. But these two possible values of the phase constant still give 2^n possible solutions of the structure, where n is the number of measurable reflections.

Each X-ray reflection is the resultant of the wavelets scattered by all the electrons in the structure, and the main problem is to discover the phase constant of this resultant wave. If a scattering particle is situated exactly at the centre of symmetry it will contribute a wavelet which is always a maximum (peak) to the resultant reflection. Now, we know that the metal atom enters the structure at the centre of symmetry (J., 1935, 615) and we assume that most of its scattering electrons are concentrated in a space which is small compared to the spacing of any of the reflecting planes. The metal atom alone, therefore, makes a contribution to the resultant reflection which is always a peak at the centre of symmetry. Thus by comparing the intensities of corresponding reflections with and without the metal atom in the structure, it is possible to derive the phase constants of the $(h0l)$ reflections. If a reflection from the free compound corresponds in phase to a peak at the centre of symmetry, then the same reflection from the metal compound will be of greater intensity owing to the addition of the peak due to the metal atom alone, and the phase constants of both these reflections will be 0 (or positive sign). If, however, the reflection from the free compound corresponds in phase to a trough at the centre of symmetry, then the addition of the peak due to the metal atom alone will tend to fill up this trough, and the resultant reflection from the metal compound will usually be of smaller intensity, the phase constants of both reflections in this case being π (or negative sign). If the reflection from the free compound corresponds to a very small trough (weak intensity), the phase may be reversed by the addition of the metal atom. When the structure amplitudes are expressed in absolute measure there is practically no ambiguity in the principal (*h*0*l*) zone, and the phase constants of these reflections, numbering about 300, from phthalocyanine and its nickel derivative have been determined with certainty. Numerical details are given below.

A similar method of determining phase constants by substitution has been successful for some inorganic structures, notably the alums (Cork, *Phil. Mag.*, 1927, **4**, 688; Beevers and Lipson, *Proc. Roy. Soc.*, 1935, A, **148**, 664), where one element can be replaced by another of higher atomic number. In the present example the results are more direct and complete, because we can remove the metal atom entirely without appreciably disturbing the structure.

The result of the Fourier synthesis for the (*h*0*l*) zone of free phthalocyanine is expressed by the contour map shown in Fig. 2, which gives a projection of the structure along the *b* crystal axis. The electron distribution is clearly segregated into peaks which there is no difficulty in identifying as due to carbon and nitrogen atoms, the latter rising to slightly higher levels and containing more electrons. Even a rough inspection of the diagram is sufficient to show that Linstead's structure for the phthalocyanines is now established with absolute certainty. Detailed measurements from the map, and the deduction of the true orientation of the molecule in the crystal, give all the interatomic distances and valency angles with an accuracy which varies slightly in different parts of the projection, but is in general about ±0·03 Å, sufficient to determine the type of valency bond between the different atoms. The results are summarised in Fig. 3, which gives a normal projection of the molecule.

Experimental

Measurement of Intensities. To determine the structure amplitudes and phase constants by the method outlined above, it was necessary to make an exhaustive series of absolute intensity measurements from free phthalocyanine and a closely isomorphous metal derivative. Nickel phthalocyanine was chosen as the most suitable because (1) the cell measurements are practically identical with those of free phthalocyanine (J., 1935, 616); (2) well-formed single crystals can be obtained; (3) the absorption coefficient for copper radiation is comparatively low, $\mu = 16\cdot56$ per cm. (for free phthalocyanine the calculated absorption coefficient is $\mu = 8\cdot62$ per cm.); (4) the scattering power of the nickel atom, atomic number 28, for X-rays is sufficient to have a marked effect on the intensities without being great enough to swamp the reflections completely. Copper phthalocyanine is almost equally good with respect to (3) and (4), and in fact, the intensities of its reflections appear to be identical with those of nickel phthalocyanine as far as can be estimated by eye, but it is not quite so suitable with respect to (1) and (2). Cu-$K\alpha$ radiation has a wavelength near the K-absorption edges of nickel and copper. The scattering powers of these elements are thus abnormal, but this does affect our results.

The crystals were in the form of small laths, elongated along the b axis. Most of the work on the ($h0l$) zone of free phthalocyanine was carried out on a specimen cut to $1\cdot02$ mm. in length (along the b axis), and of cross section $0\cdot10 \times 0\cdot34$ mm., weighing $0\cdot040$ mg. A slightly larger specimen of nickel phthalocyanine was employed, weighing $0\cdot068$ mg. The crystals were completely immersed in a uniform beam of filtered copper radiation, and the reflections recorded photographically by means of moving-film cameras. The strongest reflections were reduced by a known factor by use of automatic shutters on the two-crystal moving-film spectrometer (*Phil. Mag.*, 1934, **18**, 729). The range of intensities recorded was about 1600 to 1, and the measurements were carried out on the integrating photometer (Robinson, *J. Sci. Instr.*, 1933, **10**, 233) except for a few of the very weakest reflections, which had to be estimated visually on long-exposure films. Other crystal specimens of about the same size were employed to check the work, and for recording the other zones of reflections.

The measurements were put on an absolute scale by the direct determination of a few of the strongest reflections on the ionisation chamber with monochromatic rays (copper and molybdenum), and also photographically by comparisons with known standards on the two-crystal moving-film spectrometer. The various results were consistent to within about 5% in terms of F. The structure factors (F) were calculated from the integrated intensity measurements ($E\omega/I$), after correction for the absorption of the beam in the specimen, by the usual formula for the mosaic crystal. Adapted for a small crystal completely bathed in radiation, this becomes

$$F^2 = \frac{E\omega}{I} \cdot \frac{1}{\delta v} \cdot \frac{2\sin 2\theta}{1+\cos^2 2\theta} \cdot \frac{V^2}{\lambda^3}\left(\frac{mc^2}{e^2}\right)^2$$

δv being the volume of the specimen, and V that of the unit cell.

Owing to the shape of the specimens, the path of the X-ray beam differs when the crystals are in different reflecting positions. Correction factors were therefore calculated and applied to the comparative intensity measurements to allow for this effect. As the number of reflections is very great, this could only be done approximately, by estimating the mean path for different positions of the crystals. The results should be quite accurate for the ($h0l$) reflections, upon which the principal analysis is based, because the cross-sections of the crystals normal to the b axis are very small. For the other zones of reflections, which bring in a crystal dimension of 1 mm. or more, the corrections are less satisfactory, particularly for the nickel compound, and the resulting F values will not be so accurate; but they are quite sufficient to confirm the results of the principal analysis of the ($h0l$) zone.

Determination of Phase Constant. The unit cells contain two molecules, and therefore, in the case of nickel phthalocyanine, two nickel atoms which are situated at centres of symmetry at (000) and ($\frac{1}{2}\frac{1}{2}$0). These atoms are always in phase for the ($h0l$) zone of reflections. If the isomorphism of phthalocyanine and its nickel derivative is sufficiently close, the structure factor of the latter can be expressed as the sum or difference of two structure factors, *viz.*, that of the nickel atoms alone and that of the remainder of the compound. It is the sum if the nickel and the remainder act together, and the difference if they do not. By examining for each reflection the sum and the difference of the factors for the compound with and without nickel, it is easy to settle which gives the factor for the two nickel atoms in the cell, because the result ought to decline continuously from the full value for two nickel atoms as the value of sin θ increases. The results are set out in Table I; F (nickel phthalocyanine) and F (free phthalocyanine) may each be either positive or negative, so that four possible combinations of these quantities are involved. But as $2F$ (nickel) must always be positive, only two of these combinations need be considered, and they are given in the table.

The scattering factor for the nickel atom has recently been determined experimentally by Brindley (*Phil. Mag.*, 1935, **20**, 865; 1936, **21**, 778), but for our purpose only the very roughest knowledge of this factor is required. In fact, it is sufficient to know that the values begin at 56 (twice the atomic number) for zero glancing angle and diminish continuously as the angle increases until, for sin $\theta(\text{Cu-}K\alpha) = 0{\cdot}8$, the values of $2F$ (nickel) lie between 5 and 20, in absolute units. This approximate knowledge of the scattering factor is sufficient, because the nickel contributions calculated from the difference of the two structure factors will not, in general, be the true nickel contributions. The two structures, apart from the central nickel atom, cannot be exactly identical; but they are sufficiently alike to enable the signs or phase constants to be obtained without ambiguity for nearly all the reflections, as will be seen from the examples in Table I, where the arrow indicates the combinations of sign chosen as being correct.

About 400 equations of this type are necessary to complete the ($h0l$) zones alone for the two compounds, and only a selection of the results is given. The results are represented more completely in Fig. 1, where the calculated contributions of the nickel atoms are plotted against the sine of the 'glancing angle of the corre-

TABLE I. DETERMINATION OF PHASE CONSTANTS

hkl.	F (Ni deriv.) —	F $(C_{32}H_{18}N_8)$ =	$2F$ (Ni)	$\sin \theta$.
001	(+112·0) —	(+77·6) =	34·4 ←—	0·062
	(+112·0) —	(−77·6) =	189·6	
20$\bar{1}$	(+136·5) —	(+96·3) =	40·2 ←—	0·079
	(+136·5) —	(−96·3) =	232·8	
200	(+ 55·0) —	(A < + 3·9) =	51·1 ⎱	0·092
	(+ 55·0) —	(A < − 3·9) =	58·9 ⎰	
20$\bar{2}$	(+ 71·4) —	(+16·6) =	54·8 ←—	0·107
	(+ 71·4) —	(−16·6) =	88·0	
002	(− 36·5) —	(−85·0) =	48·5 ←—	0·123
	(+ 36·5) —	(−85·0) =	121·5	
201	(− 43·0) —	(−92·1) =	49·1 ←—	0·135
	(+ 43·0) —	(−92·1) =	135·1	
20$\bar{3}$	(+112·6) —	(+67·4) =	45·2 ←—	0·156
	(+112·6) —	(−67·4) =	180·0	
40$\bar{2}$	(− 46·0) —	(−84·8) =	38·8 ←—	0·157
	(+ 46·0) —	(−84·8) =	130·8	
40$\bar{1}$	(+129·0) —	(+77·9) =	51·1 ←—	0·159
	(+129·0) —	(−77·9) =	206·9	
400	(A < − 5·6) —	(−45·7) =	40·1 ⎱	0·183
	(A < + 5·6) —	(−45·7) =	51·3 ⎰	
310	(+ 20·4) —	(−19·2) =	39·6 ←—	0·213
	(+ 20·4) —	(+19·2) =	1·2	
004	(+ 55·6) —	(+17·0) =	38·6 ←—	0·246
	(+ 55·6) —	(−17·0) =	72·6	
20$\bar{5}$	(+ 73·5) —	(+36·0) =	37·5 ←—	0·270
	(+ 73·5) —	(−36·0) =	109·5	
60$\bar{6}$	(+ 65·7) —	(+30·7) =	35·0 ←—	0·322
	(+ 65·7) —	(−30·7) =	96·4	
40$\bar{7}$	(+ 73·1) —	(+42·4) =	30·7 ←—	0·367
	(+ 73·1) —	(−42·4) =	115·5	
024	(+ 73·5) —	(+47·8) =	25·7 ←—	0·408
	(+ 73·5) —	(−47·8) =	121·3	
12, 0$\bar{3}$	(+ 14·9) —	(−13·0) =	27·9 ←—	0·478
	(+ 14·9) —	(+13·0) =	1·9	
804	(+ 40·7) —	(+18·6) =	22·1 ←—	0·539
	(+ 40·7) —	(−18·6) =	59·3	
607	(+ 78·2) —	(+56·3) =	21·9 ←—	0·622
	(+ 78·2) —	(−56·3) =	134·5	
12, 0, $\overline{14}$	(+ 36·0) —	(+21·2) =	14·8 ←—	0·732
	(+ 36·0) —	(−21·2) =	57·2	
16, 02	(+ 23·9) —	(+15·6) =	8·3 ←—	0·805
	(+ 23·9) —	(−15·6) =	39·5	

sponding reflection. The accepted value is indicated by a dot and the rejected value by a cross. Arrows at the top of the diagram indicate further rejected values that lie beyond the limits of the scale. The reflections from the (009), (202), (80$\bar{4}$), (12,0$\bar{5}$),

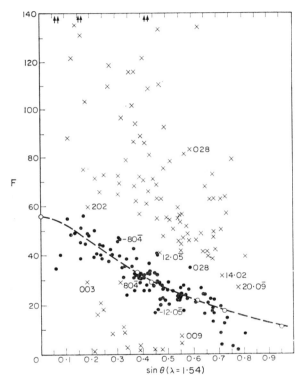

FIG. 1. The scattering contributions of the nickel atoms.

(14,02), and (20,0$\bar{9}$) planes of free phthalocyanine are extremely faint, the structure factors varying from 6 to 14 in absolute units. It will be seen from Fig. 1 that the choice of phase constants is uncertain for these reflections, but this uncertainty only applies to free phthalocyanine. In nickel phthalocyanine the (003) and the $(20,0\bar{9})$ are also faint reflections, and the corresponding signs of the structure factors are doubtful. In general, as a sum or difference of two quantities is involved, an ambiguity only arises when one of these quantities is small. This normally corresponds to an absent reflection, but those mentioned above are on the borderline of visibility. Five of these doubtful terms have been omitted from the Fourier synthesis, but they amount to such a small fraction of the total that they can have but little effect on the final results.

The broken line and circles represent the scattering power of the nickel atoms for copper radiation taken from Brindley's revised values (*Phil. Mag.*, 1936, **21**, 778). The dots are seen to congregate about this line. It should be noted that, although the dots depart considerably from a smooth curve, this does not necessarily indicate errors in the relative intensity measurements, but merely shows that the structures of free phthalocyanine and of nickel phthalocyanine, apart from the central metal atom, are not quite identical in detail.

Owing to the arrangement of the centres of symmetry in the space group C_{2h}^5, the two nickel atoms at (000) and $(\frac{1}{2}\frac{1}{2}0)$ contribute waves which are in opposite phase for certain reflections in the other zones, and consequently they make no contribution to these reflections. This happens in general when $(h+k)$ is odd, as can readily be seen from the geometrical structure factors, S (cf. "International Tables for the Determination of Crystal Structure", Vol. 1, p. 102):

$$S = 4\Sigma\cos2\pi(hx+lz)\cos2\pi ky \text{ when } (h+k) \text{ is even};$$
$$S = 4\Sigma\sin2\pi(hx+lz)\sin2\pi ky \text{ when } (h+k) \text{ is odd}.$$

Hence when $(h+k)$ is odd the structure factors should have nearly the same values for corresponding reflections from free phthalocyanine and nickel or copper phthalocyanine, a test which gives another means of checking the isomorphism of the compounds. The experimentally determined values of F for these reflections are given in Table II, and the agreements are seen

H*

TABLE II. COMPARISON OF $(h+k)$ ODD REFLECTIONS

hkl	$F,$ $C_{32}H_{18}N_8$	$F,$ Ni deriv.	hkl	$F,$ $C_{32}H_{18}N_8$	$F,$ Cu deriv.	hkl	$F,$ $C_{32}H_{18}N_8$	$F,$ Cu deriv.
011	36	33	210	53	44	11, 2, 0	36	24
012	36	35	410	21	14	13, 2, 0	$A < 14$	$A < 18$
013	$A < 11$	$A < 11$	610	10	$A < 9$	230	$A < 11$	$A < 14$
014	$A < 12$	$A < 12$	810	12	$A < 11$	430	$A < 11$	$A < 14$
015	35	35	10, 1, 0	16	$A < 13$	630	$A < 13$	$A < 15$
016	38	41	12, 1, 0	$A < 11$	$A < 15$	830	22	$A < 17$
017	23	31	120	$A < 7$	$A < 10$	10, 3, 0	11	$A < 20$
018	$A < 19$	$A < 19$	320	$A < 8$	$A < 10$	140		$A < 16$
019	$A < 20$	$A < 20$	520	32	26	340	$A < 12$	$A < 16$
031	$A < 17$	$A < 17$	720	$A < 10$	$A < 13$	540	$A < 14$	$A < 18$
032	$A < 17$	$A < 17$	920	$A < 11$	$A < 14$	740	16	$A < 18$
033	16	14						

TABLE III. VALUES AND SIGNS OF $F(h0l)$

					h					
l	0	2	4	6	8	10	12	14	16	18
11										
10	+13	−19		+16						
9		−42.5		+56.5						
8	−28	−14	+10	+32		−12				
7	+23.5	+15	+35.5	+29	+26					
6	50	+24.5	−13	+26	+18.5	−17.5	−35			
5	−59.5	−72	−66.5	−33		−19				
4	+17	+9.5		−75.5		+13	−11			
3	39	−13	−12.5	−37	+21.5	−25				
2	+85	+5.5	−45.5	+38		+11.5	+16	+14	+15.5	
1	+77.5	−92	+80	−40	+20	+61.5	+12.5	−32.5	−10	
0	+532		−85	−55		+20	−13		−15	
1	+77.5	+96.5	−74	−30.5	+15	+28	+52		−12.5	
2	−85	+16.5	−48	−16		+12.5	+11.5		+10	
3	39	+67.5	−37.5	+30.5	−13	−13.5	−41			
4	−17	−60.5		+30.5		+27	−30.5			
5	+59.5	+36	+42.5	+16.5	−14.5		−42	−29.5	−28.5	
6	50		+58.5	−22	+41.5	−46	−13	−9.5		−26.5
7	+23.5	+21.5	+12.5		+44	+14.5	+10	+11		+25.5
8	28		+16.5		−15			+16		+27
9				−24.5	−38		+8.5	+9	+8.5	
10	+13	−38	−24	−27		−12.5		−22	+9.5	
11			−16		+27	+25	+21	−36.5	+10	
12		+38			−18					
13										
14										

to be reasonably good. It follows, of course, that the phase constants of these reflections cannot be determined directly as in the case of the (*h*0*l*) zone. Fortunately, however, the latter zone is sufficient for a complete deduction of the structure, and from the results obtained, the magnitudes and phase constants of the reflections in Table II can be calculated and used to verify the main conclusions. The calculated values of the structure factors will be given in a later paper when the structure of nickel phthalocyanine is considered in detail.

Fourier Analysis of the (h0l) *Zone, and Orientation of the Molecule.* The values and signs of $F(h0l)$ are given in Table III. The projected density was computed from the series

$$\rho(x,z) = \frac{1}{ac \sin \beta} \sum_{-\infty}^{+\infty} \sum_{-\infty}^{+\infty} F(h0l) \cos 2\pi(hx/a + lz/c)$$

The summations were carried out at 1800 separate points on the asymmetric unit (half the molecule), the *a* axis being sub-divided into 120 parts (intervals of 0·165 Å) and the *c* axis into 60 parts (intervals of 0·247 Å). The summation totals are not reproduced, but the results are represented by the contour map of Fig. 2, which is drawn by interpolation methods from the totals. It may be noted that the summations involve the addition of over 230,000 terms, but the numerical work can be considerably reduced by methods already described (*Phil. Mag.*, 1936, **21**, 176).

All the atoms in the molecule (except hydrogen) can be located without difficulty in the projection, and the positions of their centres can be determined with some accuracy. Hence, two co-ordinates (*x* and *z*) for each atom can be obtained directly, but the third co-ordinate (*y*), which is measured normal to the plane of the projection, is unknown. But the remarkable regularity of the structure makes it easy to estimate the *y* co-ordinates, and the results can later be checked against another projection of the structure described below.

It is convenient to refer the measurements to two principal molecular axes L and M (Fig. 2). When lines are drawn to connect opposite pairs of atoms, (11, 11′), (10, 12′), (9, 13′), (4, 18′), etc., it is found that these lines are all parallel to within $\pm0.35°$. The mean direction of 19 such lines is taken as the M axis, and this makes an angle of 22·9° with the a crystal axis

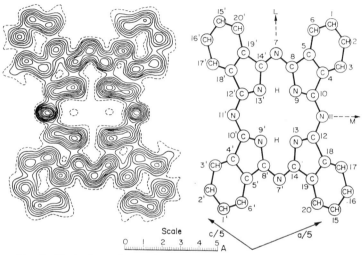

Fig. 2. Projection along the b axis, showing one complete phthalocyanine molecule. The plane of the molecule is steeply inclined to the plane of the projection, the M direction making an angle of 46° with the b axis, and the L direction 2·3°. Each contour line represents a density increment of one electron per Å², the one-electron line being dotted.

(η_M). Similarly, lines connecting pairs of atoms above and below the centre (7, 7′), (8, 14), (9, 13), etc., are also found to be parallel to within $\pm0.4°$, and their mean direction (L) makes an angle of 69·3° with the a axis (η_L).

The only simple explanation of this extreme regularity in an arbitary projection of the structure is that the whole molecule is planar. If the 38 lines mentioned above were not parallel in the actual molecule, it would be an extraordinary coincidence that they are parallel in this projection. This does not exclude the possibility of troughs running across the molecule, but the regularity of the distances (see below) and the presence of the centre of symmetry make this very unlikely.

The benzene rings at the four corners of the molecule are found to be the projections of regular plane hexagons, to within about 0·04 Å; *i.e.*, opposite sides are parallel to, and of one-half the length of, the line through the centre joining the other two corners. Now the size of the benzene ring has frequently been measured in organic compounds (*Proc. Roy. Soc.*, 1933, **142**, 659, etc), and the most reliable value for the radius, or centre to centre distance, is 1·39 Å. If we take this figure as a known dimension, the four benzene rings in the structure can then be used as "indicators", to determine the true orientation of the molecule in the crystal. For example, the benzene hexagons are greatly foreshortened in the *M* direction, indicating a considerable inclination of that axis of the molecule to the plane of the projection. Along the *L* axis the distances are practically those of normal rings, showing that the inclination of this axis to the plane of the projection must be small (less than 10°). Two of the four benzene rings are crystallographically independent, and the other two are inversions of these through the centre of symmetry. But as all four were drawn by independent interpolation from the summation totals, the results are given separately. The angle between the *M* axis and the *b* crystal axis is donated by ψ_M and we have $\sin\psi_M = r/R$, where *r* is the measured length of a line in the projection, and *R* is the real length (inferred). In this way we obtain the following values for ψ_M from the four rings

Ring 1 ... 4 $\psi_M = 46\cdot6°$ Ring 15 ... 18 $\psi_M = 45\cdot0°$
,,   1' ... 4' $\psi_M = 46\cdot3$,,   15' ... 18' $\psi_M = 46\cdot2$
Mean $\psi_M = 45\cdot9°$.

The mean result for the inclination of the *M* axis of the molecule is probably accurate to within 1°. The inclination of the *L* axis, however, cannot be determined in this way, because when r/R is near unity the results are unreliable. But another method is available. The two molecular axes L and M are inclined at 92·2° to each other in the projection, $(\eta_L + \eta_M)$, a figure which is probably accurate to within 0·2°, as it is the mean of 19 different pairs of lines. Now it is safe to assume that these two axes are at right angles in the actual molecule. If they were not, interatomic distances on one side of the molecule, including distances within the benzene ring, would be contracted, while corresponding distances on the other side would be relatively expanded, a condition which is extremely improbable. The apparent departure from perpendicularity of the axes L and M must be due to the orientation of the molecule and can therefore be used to give an accurate measure of the inclination of the L axis from the plane of the projection. We now have the necessary conditions for deriving the complete orientation of the molecule in the crystal. χ_L, ψ_L, and ω_L denote the real angles which the L axis makes with the a and b crystal axes and their perpendicular, while χ_M, ψ_M, and ω_M are the corresponding angles for the M axis. η_L and η_M are the observed angles which L and M make with the a axis in the projection. The conditions are

(1) $\cos^2\chi_L + \cos^2\psi_L + \cos^2\omega_L = 1$
(2) $\cos^2\chi_M + \cos^2\psi_M + \cos^2\omega_M = 1$
(3) $\cos\chi_L\cos\chi_M + \cos\psi_L\cos\psi_M + \cos\omega_L\cos\omega_M = 0$
(4) $\cos\omega_L = \cos\chi_L\tan\eta_L$
(5) $\cos\omega_M = \cos\chi_M\tan\eta_M$
(6) $\psi_M = 45·9°$

from which we obtain the following values for the direction cosines of the molecular axes

$\chi_L = 69·3°, \cos\chi_L = 0·3532 \quad \chi_M = 48·6°, \cos\chi_M = 0·6615$
$\psi_L = 87·7°, \cos\psi_L = 0·0395 \quad \psi_M = 45·9°, \cos\psi_M = 0·6960$
$\omega_L = 20·8°, \cos\omega_L = 0·9347 \quad \omega_M = 106·2°, \cos\omega_M = -0·2794$

$$\chi_N = 131·4°, \cos\chi_N = -0·6617$$
$$\psi_N = 44·2°, \cos\psi_N = 0·7169$$
$$\omega_N = 77·3°, \cos\omega_N = 0·2197$$

The last three quantities refer to the direction of the normal to the molecular plane.

Molecular Dimensions and Co-ordinates.   Although all the carbon and nitrogen atoms in the molecule can be clearly seen in the projection of Fig. 2, the precision with which their centres can be determined necessarily varies in different parts of the

TABLE IV.   Co-ordinates With Respect to Molecular Axes

Atoms	*l* measured, Å	*m* measured, Å	Mean value, Å L ($l \sin \psi_L$)	M ($m \sin \psi_M$)
1, 15	5·05, 5·09	2·99, 3·04	±5·07	4·20
2, 16	4·05, 4·10	3·70, 3·72	±4·08	5·17
3, 17	2·72, 2·78	3·43, 3·42	±2·75	4·77
4, 18	2·39, 2·43	2·45, 2·46	±2·41	3·42
5, 19	3·39, 3·42	1·75, 1·77	±3·40$_5$	2·45
6, 20	4·71, 4·75	2·02, 2·06	±4·73	2·84
7	3·43	0	±3·43	0
8, 14	2·70, 2·72	0·81, 0·81	±2·71	1·13
9, 13	1·39, 1·38	0·94, 0·96	±1·38	1·32$_5$
10, 12	1·13, 1·15	1·90, 1·89	±1·14$_5$	2·64
11	0	2·38	0	3·32

projection, and can best be estimated by a study of the map. The nitrogen atoms 11 and 7 are most clearly defined, and in the benzene ring, atoms 2, 5, 16, and 19. The centres of these can be determined fairly accurately (± 0.02 Å), but the resolution of the other atoms is not quite so satisfactory. It is found, however, that projections of regular plane hexagons can be inscribed on the benzene rings, and that the atoms lie on the corners of these hexagons, with some possible small deviations in the case of the less perfectly resolved atoms, which are noted in the next section. The positions of the centres of all the atoms were estimated, and the co-ordinates measured parallel to the molecular axes L and M, with the results shown in Table IV.

The atoms are grouped in symmetrical pairs above and below the M axis, and are denoted by the numbers in the first column.

Individual measurements of the co-ordinates of these atoms are given by the next four numbers. The four-fold symmetry of the molecule is shown by the similarity of the co-ordinates of the two atoms in each pair. The differences are so small that they may be ascribed to experimental errors, and consequently the numbers have been averaged in the last two columns, and multiplied by the sines of the inclinations of the molecular axes to

TABLE V. CO-ORDINATES WITH RESPECT TO MONOCLINIC CRYSTAL AXES. CENTRE OF SYMMETRY AS ORIGIN

Atom (cf. Fig. 2)	x, Å	$2\pi x/a$	y, Å	$2\pi y/b$	z, Å	$2\pi z/b$
1 CH	6·82	123·7°	3·12	238·1°	4·22	102·7°
2 CH	6·35	115·2	3·76	286·5	2·80	68·1
3 CH	4·91	89·0	3·43	261·5	1·46	35·6
4 C	3·94	71·3	2·48	189·0	1·53	37·3
5 C	4·40	79·8	1·84	140·6	2·95	71·8
6 CH	5·84	105·8	2·17	165·2	4·29	104·3
7 N	3·24	58·7	0·14	10·4	3·79	92·2
8 C	3·10	56·3	0·90	68·2	2·62	63·8
9 N	1·95	35·2	0·98	74·5	1·09	26·5
10 C	2·36	42·8	1·88	143·6	0·39	9·6
11 N	1·61	29·1	2·31	176·0	− 1·10	− 26·7
12 C	0·20	3·7	1·79	136·7	−2·14	− 52·0
13 N	−0·66	− 12·0	0·87	66·1	−1·96	− 47·8
14 C	−2·01	− 36·4	0·68	52·0	−3·37	− 81·9
15 CH	−2·76	− 50·0	2·72	207·6	−6·99	−170·1
16 CH	−1·34	− 24·3	3·43	262·0	−6·21	−151·2
17 CH	−0·28	− 5·0	3·21	245·0	−4·61	−112·2
18 C	−0·61	− 11·1	2·29	174·5	−3·80	− 92·3
19 C	−2·02	− 36·6	1·57	119·9	−4·57	−111·2
20 CH	−3·08	− 55·8	1·79	136·6	−6·16	−150·0

the projection axis (b), which reduces the measured values to actual molecular dimensions. From these figures the interatomic distances and valency angles are easily obtained, and the results are shown in Fig. 3. There is a possibility that atoms 8 and 14′ may lie about 0·04 Å nearer to 7 than is shown in Fig. 3. The positions assigned, however make allowance for overlap, and seem the most probable.

The co-ordinates of the atoms referred to the crystal axes, which are required for calculating the structure factors, etc., are obtained by combining the molecular co-ordinates L and M with the orientation angles according to the relations

$$x' = L \cos \chi_L + M \cos \chi_M$$
$$y = L \cos \psi_L + M \cos \psi_M \qquad x = x' - z' \cot \beta$$
$$z' = L \cos \omega_L + M \cos \omega_M \qquad z = z' \operatorname{cosec} \beta$$

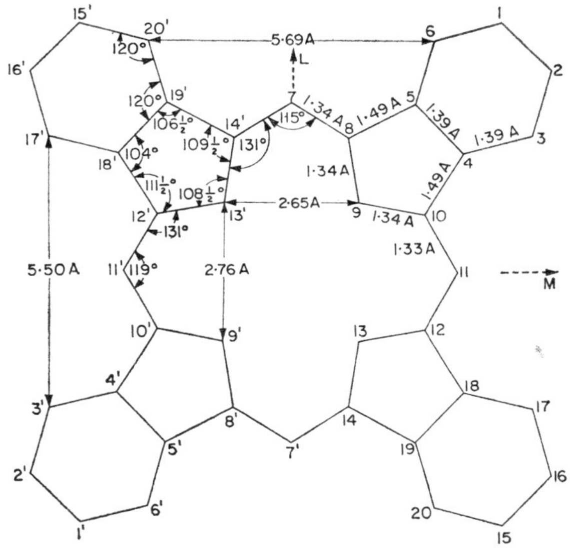

Dimensions of the phthalocyanine molecule

FIG. 3. Dimensions of the phthalocyanine molecule.

$(x'yz')$ are rectangular co-ordinates referred to the a and b crystal axes and their perpendicular, and (xyz) are the monoclinic crystal co-ordinates. The numerical results are given in Table V. It should be noted that x and z can be obtained by direct measurement of the projection, but y can only be obtained by calculation.

Discussion of the Structure

The molecular dimensions shown in Fig. 3 indicate the type of bond between the various atoms. The closed inner system consisting of 16 carbon and nitrogen atoms is one of great stability, and is of frequent occurrence in Nature, forming the nucleus of the porphyrins. The interatomic distance has the practically constant value of 1·34 Å, with an average possible error of about ±0·03 Å, a figure which points to single bond–double bond resonance between the carbon and nitrogen atoms.

This inner system is connected to the benzene rings by carbon–carbon bonds of quite a different type, the distance here being 1·49 ± 0·03 Å. If we use the empirical function expressing the dependence of carbon–carbon interatomic distance on bond character for single bond–double bond resonance given by Pauling, Brockway, and Beach (*J. Amer. Chem. Soc.*, 1935, **57**, 2706), this distance of 1·49 Å would indicate from 12 to 15% of double-bond character in these links. This is very similar to the type of bond between conjugated benzene rings in diphenyl and *p*-diphenylbenzene (Dhar, *Indian J. Physics*, 1932, **7**, 43; Pickett, *Proc. Roy. Soc.*, 1933, A, **142**, 333). From the chemical structure (I), however, we should expect the somewhat higher double-bond character of 25%; but the main point of interest lies in the equality, within narrow limits, of these eight bonds which connect the benzene rings to the inner resonating system. This result emerges directly from the X-ray measurements.

(I.)   (II.)   (III.)

With regard to the benzene rings, the dimension of 1·39 Å was assumed in calculating the orientation of the molecules; but the fact that the rings appear as sensibly *regular* hexagons confirms the validity of this assumption.   Some of the carbon atoms (*e.g.*, 1, 4, 6, 20) are not very perfectly resolved, and in order to see if any systematic deviations from regularity might exist, an attempt was made to estimate the centres of all the atoms independently, without trying to inscribe the projections of regular plane hexagons on the rings. The interatomic distances obtained were as follows (in Å):

1, 2 = 1·39	4, 5 = 1·36	15, 16 = 1·35	18, 19 = 1·37
2, 3 = 1·40	5, 6 = 1·38	16, 17 = 1·39	19, 20 = 1·38
3, 4 = 1·44	6, 1 = 1·35	17, 18 = 1·38	20, 15 = 1·35

The angles were all within 3° of 120°. On the whole, systematic deviations do not seem to occur. The variations are, perhaps, not larger than might be expected from the lack of resolution of the atoms, due to incompleteness of the Fourier series and experimental errors.

There are two conflicting possibilities, either of which might affect the dimensions of the benzene rings in this structure. If the links connecting them to the inner system have sufficiently high double-bond character, there might be a tendency to *o*-quinonoid ring formation, as in (II). On the other hand, measurement of the angles in the *iso*indole ring (compare Fig. 3) shows that the links emerging from the benzene rings are very considerably strained out of the normal (120°) positions, the mean value of the two angles being about 105°. Following the work of Mills and Nixon (J., 1930, 2510), and more recently of Sidgwick and Springall (this vol., in the press), this might lead us to expect a different stabilisation of the Kekulé forms, as shown in (III). In both (II) and (III) the 4, 5 link might be expected to exhibit a certain degree of single-bond character beyond the other links, but there is no experimental evidence of this in our results.

There does exist, however, another distortion, affecting the whole molecule, which appears to be large enough to permit of fairly definite measurement. This is a departure from true tetragonal symmetry, and is best seen by comparing the lengths of the horizontal and vertical lines joining corresponding pairs of atoms (compare Fig. 3 and Table IV). Taking first the lines connecting atoms on the inner carbon-nitrogen system, the lengths (in Å) are

Horizontal lines.		Vertical lines.	
8, 14′ = 2·26	10, 12′ = 5·28	10, 12 = 2·29	8, 14 = 5·42
9, 13′ = 2·65	11, 11′ = 6·64	9, 13 = 2·76	7, 7′ = 6·86

The horizontal lines are all shorter than the corresponding vertical lines, the discrepancy increasing as we descend from the apex of the molecule at the nitrogen atom 7. If, however, the lines are taken to connect corresponding atoms on the benzene rings, the situation is reversed, the *horizontal* dimensions in Fig. 3 being the greater, as shown below.

Horizontal lines.		Vertical lines.	
1, 15′ = 8·40	5, 19′ = 4·90	2, 16 = 8·16	4, 18 = 4·82
6, 20′ = 5·69	3, 17′ = 9·54	3, 17 = 5·50	6, 20 = 9·46
2, 16′ = 10·34	4, 18′ = 6·84	1, 15 = 10·14	5, 19 = 6·81

In this case the discrepancy diminishes as we descend from the highest line, 1, 15′.

It is impossible to explain these two contrary departures from true tetragonal symmetry by an error in calculating the orientation of the molecule, and the differences seem much too systematic to be accounted for by random experimental errors, or deficiencies in the Fourier series. They can be explained very easily by the simplified model shown in Fig. 4, where the zig-zag C–N links (7, 8, 9, 10, 11) are replaced by the straight line *AB*. The benzene ring is situated on the outer portion of the line *CD*. If a distortion is caused, *e.g.*, by a force operating between *E* and *F*, the dotted lines show that the dimensions change in a manner which explains all the measurements given above, at least qualitatively. The angle at *A* will diminish, and that at *B* will increase. In the actual molecule these angles are 115° and 119°.

EF in the simple model corresponds to the distance between the *iso*indole nitrogen atoms (9, 13′) in the molecule, which is 2·65 Å. Now this distance is just about the right order for hydrogen-bond formation as between oxygen atoms in acids and acid salts (Pauling, *J. Amer. Chem. Soc.*, 1931, **53**, 1367, etc.). Quantitative data on the distances involved in hydrogen-bond formation between nitrogen atoms are not available, but there seems little doubt that they can occur. The presence of such *internal* hydrogen bonds in the phthalocyanine molecule would help to explain the distortion from true tetragonal symmetry.

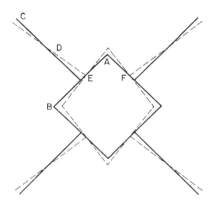

FIG. 4. Simplified model of the phthalocyanine molecule.

Another Projection of the Structure. No other projection of the structure is possible which gives a resolution of the atoms in any way approaching that obtained along the *b* axis. If we try to make a projection nearly normal to the plane of the molecule, other interleaving molecules come behind, and the resulting diagram is hopelessly obscured. It is, however, important to study the other zones of reflections to obtain more direct confirmation of the values assigned to the *y* co-ordinates, which are based upon the calculated orientation and assumed flatness of the molecule. When the structure factors for these other zones are calculated from the co-ordinates in Table V, good agreements with the observed values are obtained, of which details will be given in a later paper. The projection along the c axis, corresponding to the $(hk0)$ zone of reflections, has been chosen for detailed study, and a Fourier synthesis is given in Fig. 5. The coefficients are given in Table VI. The signs, or phase constants, of the $(h+k)$ even planes were obtained in the same direct manner as before, but for the $(h+k)$ odd planes the signs are from the calculated values of the structure factors.

The drawing to the right of the contour map is prepared from the co-ordinates in Table V (atoms half size) and shows the complicated way in which seven phthalocyanine molecules overlap in this projection. Only the portion within the dotted line, representing one unit cell, is complete.

Only the atoms 3 and 3' are separately resolved in this projection of the structure, but a careful study of the map shows that the contour lines correspond faithfully to the calculated positions of the atoms. This is especially evident near the empty portions of the structure at the centre, and at the centres of the benzene rings. If any considerable distortion from the planar form existed in the molecule, it would be evident from this projection.

The co-ordinates of atom 3 estimated from this projection compare with the previous values as follows:

	x, A.	y, A.
Measured from c projection	4·78	3·34
From b projection and calc. orientation....	4·91	3·43

The agreements are perhaps as close as can be expected, in view of the unsatisfactory nature of this projection, and the difficulty of making accurate intensity measurements in this zone.

Intermolecular Distances and Arrangement of Molecules in the Crystal. The way in which the phthalocyanine molecules interlock to form the crystal is shown by the group in Fig. 6, which is a small-scale reproduction of the b-axis projection. Although appearing identical in this projection, the planes of the molecules are actually inclined in opposite directions to the plane of the

TABLE VI. VALUES AND SIGNS OF F($hk0$)

		k				
		0	1	2	3	4
	16	− 15	—	—	—	—
	15	—	—	—	—	—
	14	—	—	+17·5	—	—
	13	—	—	—	—	—
	12	+ 16	—	−12	—	—
	11	—	—	+36	—	—
	10	+11·5	+16	—	−22·5	—
h	9	—	−28	—	—	—
	8	—	−12·5	−10	—	—
	7	—	—	—	—	−16
	6	+ 38	+10	+14·5	—	—
	5	—	—	+31·5	+13	—
	4	− 45·5	+20·5	—	—	—
	3	—	−19	—	—	—
	2	—	−52·5	−10·5	—	—
	1	—	−65·5	—	−14·5	+10·5
	0	+532	—	−10	—	+ 6
	$\bar{1}$	—	−65·5	—	−14·5	−10·5
	$\bar{2}$	—	+52·5	−10·5	—	—
	$\bar{3}$	—	−19	—	—	—
	$\bar{4}$	−45·5	−20·5	—	—	—
	$\bar{5}$	—	—	−31·5	+13	—
	$\bar{6}$	+38	−10	+14·5	—	—
	$\bar{7}$	—	—	—	—	+16
	$\bar{8}$	—	+12·5	−10	—	—
h	$\bar{9}$	—	−28	—	—	—
	$\overline{10}$	+11·5	−16	—	+22·5	—
	$\overline{11}$	—	—	−36	—	—
	$\overline{12}$	+16	—	−12	—	—
	$\overline{13}$	—	—	—	—	—
	$\overline{14}$	—	—	+17·5	—	—
	$\overline{15}$	—	—	—	—	—
	$\overline{16}$	−15	—	—	—	—

projection (010). Thus, if the two end molecules in the lower row are inclined to the right, the middle one is inclined to the left, and the two end ones of the upper row are inclined to the left.

The molecular planes are inclined at 44·2° to the (010) plane, and consequently the planes of adjacent molecules are almost perpendicular.

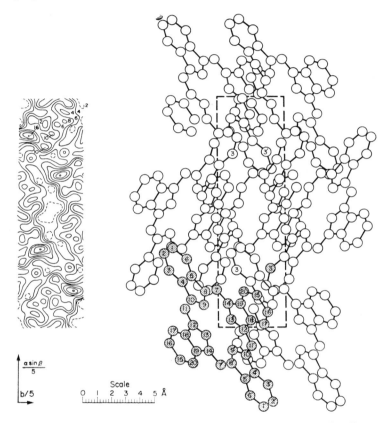

FIG. 5. Projection along the *c* axis showing the complete unit cell, and how parts of seven phthalocyanine molecules contribute to the structure. Each contour line represents 2 electrons per Å², the two-electron line being dotted.

Identical molecules recur along the *b* axis at intervals of 4·72 Å, and the perpendicular distance between the planes of these parallel molecules is $b.\cos\psi_N = 3\cdot38$ Å, almost identical with the interlaminar distance in graphite (3·41 Å). When a projection is taken normal to the molecular planes, none of the atoms in the parallel molecules at either end of the b axis comes exactly over another, the nearest approaches being between 17 and 14,

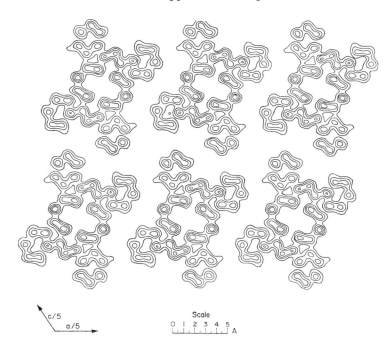

c/5

a/5

Scale
0 1 2 3 4 5 A

Fig. 6. A group of phthalocyanine molecules in the b axis projection. Each contour line represents two electrons per Å².

and between 3 and 8 (carbon atoms) where the distances are 3·41 Å. The nitrogen atoms 13 and 9 are nearly as close, the distance being 3·44 Å.

The shortest intermolecular distance which has been found in the whole structure occurs between the benzene carbon atom

2 on the standard molecule, and the nitrogen atom 11' on the reflected molecule ($\frac{1}{2}a$, $\frac{1}{2}b$, removed) where the value is 3·35 Å. No other approach of less than 3·5 Å has been found, although between the benzene carbon 1 on the standard molecule and the benzene carbon 15 on the reflected molecule one translation along the c axis and one translation along the b axis, the distance is as low as 3·63 Å. These results show that the usual order of minimum intermolecular distance, which occurs when only van der Waals forces are operative between the molecules, is maintained in the phthalocyanine structure.

Summary

The structure of phthalocyanine has been determined by direct X-ray analysis of the crystal. The usual difficulty of the unknown phase constant, which necessitates a preliminary analysis by trial, has been overcome by comparing absolute measurements of corresponding reflections from nickel phthalocyanine and the metal-free compound, which leads to a direct determination of all the significant phase constants in the ($h0l$) zones of the two compounds, numbering about 300. A Fourier analysis of these results determines two co-ordinates of each carbon and nitrogen atom in the structure, and the regularity of the projection shows beyond doubt that the molecule is planar. The orientation of the molecule in the crystal is deduced, and the third co-ordinates of the atoms calculated. The results are then confirmed by a second Fourier projection of the structure, along the c axis.

The results show that the chemical structure assigned to the compound by Linstead is correct, and give further information about the type of valency bonds. The inner nucleus of the molecule, which is common to the porphyrins, consists of a closed system of 16 carbon and nitrogen atoms, which appear to be in a state of double bond–single bond resonance, the interatomic distance having the appreciably constant value of $1·34 \pm 0·03$ Å. This inner system is connected to the four benzene rings by C–C bonds of length $1·49 \pm 0·03$ Å, which indicates a small percentage of double-bond character.   The benzene rings are sensibly regular, such variations in their C–C distance as do occur being of small order and apparently erratic, probably owing to deficiencies in the Fourier series.   The carbon links emerging from the benzene rings are strained about 15° from their normal positions. In addition, the molecule as a whole suffers a small distortion from true tetragonal symmetry, probably due to the existence of an internal hydrogen bond between the two *iso*indole nitrogen atoms.

In the crystal, the planes of adjoining rows of molecules are almost at right angles, and the minimum intermolecular approach distance is 3·35 Å.   The perpendicular distance between the molecular planes of parallel molecules is 3·38 Å.

The Davy Faraday Research Laboratory of the Royal
Institution,

Received June 26th, 1936.

10 A Fourier Series Method for the Determination of the Components of Interatomic Distances in Crystals*

A. L. PATTERSON†

A method for the direct determination of the components of interatomic distances in crystals has been developed from a consideration of the properties of the Fourier series whose coefficients are the squares of the F-coefficients for the crystal reflections. Valuable structural information is thus obtained without making any assumptions as to the phase to be allotted to the F-coefficients. The practical application of the method is illustrated by a discussion of the structures of potassium dihydrogen phosphate and hexachlorobenzene.

1. Introduction

In any crystal, the density of scattering power for X-rays (electron density) can be represented by a three-dimensional Fourier series of the form[1]

$$\rho(xyz) = \sum_{hkl=-\infty}^{\infty} \sum \sum \alpha(hkl)e^{2\pi i(hx/a + ky/b + lz/c)}. \tag{1}$$

If this density is real, as is usually the case, we have, in addition,

$$\alpha(h, k, l) = \alpha^*(-h, -k, -l), \tag{1a}$$

*Physical Review, 46, 372–6 (1934). Presented in part at the Washington Meeting of the American Physical Society, Phys. Rev. 45, 763A (1934).

†George Eastman Research Laboratories, Massachusetts Institute of Technology. Received June 18, 1934.

[1]For literature references and notation see W. L. Bragg, Proc. Roy. Soc. A 123, 537 (1929); also A. L. Patterson, Zeits. f. Krist. 76, 177 (1930).

where α^* is the conjugate complex of α. It is a well-known result that the values of $F(hkl)$ obtained from absolute measurements of integrated intensity of X-ray reflection are connected with the coefficients $\alpha(hkl)$ of the series (1) by the relation

$$F(hkl) = |\alpha(hkl)|. \qquad (2)$$

The problem of X-ray crystal analysis is the determination of the appropriate phases for the quantities $\alpha(hkl)$. This involves the use of our knowledge of the atomic scattering powers of the atoms of which the crystal is composed. These atoms are allotted positions in the unit cell in accordance with the space group requirements. Each of these positions, except in very special cases, involves one or more parameters in its specification. In general, a crystal structure investigation will involve the determination of a large number of such parameters and their calculation in most cases can only be carried out by a process of trial and error.

In this paper a method is presented which enables the principal interatomic distances to be directly determined. The directions in which these distances lie can also be obtained. No assumptions are involved in the deduction of these results and they are independent of the space group determination.

These interatomic distances place very definite limits on the values which the unknown parameters can assume. The labor involved in their determination is thus very considerably reduced.

2. One-Dimensional Problem

It is simpler to discuss the one-dimensional problem first. Consider the distribution of electron density normal to a crystallographic plane whose spacing is d. This density can be expressed in the form

$$\rho(x) = \sum_{-\infty}^{\infty} \alpha(n)e^{2\pi inx/a}; \quad \alpha(n) = \alpha^*(-n). \qquad (3)$$

Let the curve of Fig. 1 represent any such distribution function. Consider an element dx at a distance x from the origin. The distribution around this element can be expressed as a function of a parameter t in the form $\rho(x+t)$. Suppose we weight this distribution by the quantity $\rho(x)dx$, the amount of scattering matter in the element dx, and compute the weighted average distribution about any element dx when x is allowed to assume all values within the period. This average distribution $A(t)$ is given by

$$A(t) = d^{-1} \int_0^d \rho(x)\rho(x+t)dx. \qquad (4)$$

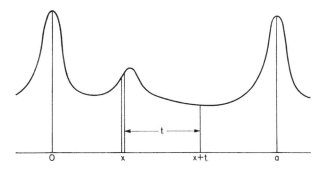

Fig. 1.

This integral is well known in the modern theory of the Fourier series as the "Faltung" of $\rho(x)$ and can immediately be evaluated by substituting (3) in (4), i.e.,

$$A(t) = \sum \left| \alpha(n) \right|^2 e^{2\pi int/d}. \qquad (5)$$

This series can obviously be computed directly from the measured intensities of reflection from the various orders of the plane and it will be of great value to crystal analysis if we can obtain a simple physical interpretation of its meaning. This is readily done for the integral form, as follows.

We see that the principal contributions to $A(t)$ will be made when both $\rho(x)$ and $\rho(x+t)$ have large values. Thus if there is a peak in the curve $A(t)$ for a value $t = t_1$, it means simply that there are two peaks in the curve $\rho(x)$ at a distance t_1 apart. This simple qualitative result has a direct interpretation in crystal analysis. If we find a peak in the curve $A(t)$ for a given plane at a distance t_1 from the origin, we know that somewhere in the distribution normal to the plane there are planes of atoms at a distance t_1 apart.[2]

While this qualitative result is very simple, its exact mathematical expression is extremely difficult. The distances between peaks are not reproduced exactly except in the simplest cases and that makes a strict mathematical interpretation almost impossible. However, a trial with a few simple series will convince the reader that the accuracy is sufficiently good to give very useful approximations for these distances. This is also shown by the examples given below.

3. *Three-Dimensional Series*

The result of the preceding paragraph can be extended immediately to three dimensions. We evaluate the integral

$$A(uvw) = (abc)^{-1} \int_0^a \int_0^b \int_0^c \rho(xyz) \times$$

$$\times \rho(x+u, y+v, z+w)dxdydz$$

$$= \sum \sum \sum |\alpha(hkl)|^2 e^{2\pi i(hu/a + kv/b + lw/c)}$$

$$= \sum \sum \sum F^2(hkl)e^{2\pi i(hu/a + kv/b + lw/c)}. \tag{6}$$

By a direct extension of the above argument, we can show that if

---

[2] The result obtained here is an extension of the application to crystals of the theory of scattering of X-rays in liquids reported by Gingrich and Warren at the Washington meeting of the American Physical Society and arose in a discussion of that work. *Phys. Rev.* **46**, 368 (1934).

we find a maximum of A (uvw) at some point $(u_1v_1w_1)$, then there are two maxima (atoms) in the distribution $\rho(xyz)$ whose distance apart is given by the vector whose components are $(u_1v_1w_1)$.

4. *Two-Dimensional Series*

In the practical application of this method to the analysis of crystals, the two-dimensional series promises to be the most fruitful. It is much easier to compute and much easier to represent than the three-dimensional series; and it is much less confused and more easily interpreted than the one-dimensional series. We consider a distribution of the type

$$A\ (uv) = \sum_{hk=-\infty}^{\infty}\sum F^2(hk0)e^{2\pi i(hu/a + kv/b)} \tag{7}$$

and discuss the components of the interatomic distances which lie in the plane under consideration.

5. *Examples of the Method*

The practical application of the method is best discussed in the light of known structures. Two such examples have been chosen. Potassium dihydrogen phosphate, which has been very thoroughly investigated by West[3] provides an example of a simple inorganic substance with several atoms in fixed positions and one set of atoms in a general position involving three parameters. A complete set of absolute measurements of intensity is available for the two principal zones of the crystal. The second example is hexachlorobenzene,[4] a relatively simple organic structure whose atoms are all in general positions. For this crystal, Lonsdale has obtained a satisfactory set of relative intensities for the zone [010] from which a picture is obtained of the projection on the (010) face. No original data have been obtained for either of these crystals. The present paper merely

---

[3] J. West, *Zeits. f. Krist.* **74**, 306 (1930).
[4] K. Lonsdale, *Proc. Roy. Soc.* A **133**, 536 (1931).

involves a rediscussion of the published data, making use of the new method of analysis.

(a) *Potassium dihydrogen phosphate.* This substance crystallizes in the space group $V_d^{12}(I\bar{4}2d)$. The unit cell, whose dimensions are $a = b = 7.43\text{Å}$; $c = 6.97\text{Å}$, contains four molecules KH_2PO_4. The positions of the potassium and phosphorus atoms are fixed by symmetry conditions. Qualitative consideration of the complexity of the spectra indicates that the sixteen oxygen atoms occupy the sixteen-fold general position[5] requiring three parameters for its specification. We shall confine ourselves here to the application of the new method to the determination of the x and y parameters of the oxygen atoms.

We compute[6] the series

$$\sum \sum F^2(hk0)e^{2\pi i(hu/a + kv/b)},$$

making use of the observed absolute F values published by West. The result of this computation is shown in the form of a contour map in Fig. 2(*b*). From the positions of the eight peaks surrounding the origin and our knowledge of the space group, we can immediately determine the oxygen parameters in this plane. The maxima occur for the values[7] $\theta_1 = 29.6°$ and $\theta_2 = 51.4°$. These are to be compared with the values obtained by West, i.e., $\theta_1 = 29°$, $\theta_2 = 52°$.

(b) *Hexachlorobenzene.* The space group in this case is $C_{2h}^5(P2_1/c)$. There are two molecules C_6Cl_6 in a cell of dimensions $a = 8.07\text{Å}$, $b = 3.84\text{Å}$, $c = 16.61\text{Å}$, $\beta = 116°25'$. All the atoms are in twofold general positions,[5] the molecules having central symmetry.

[5] R. W. G. Wyckoff, *Analytical Expression of the Results of the Theory of Space Groups*, Washington, 1930.
[6] The method of Beevers and Lipson (*Phil. Mag.* **17**, 855 (1934)) is very convenient.
[7] $\theta_1 = 360u/a$, etc.

In this case we compute the series

$$\sum \sum F^2(h0l)e^{2\pi i(hu/a+lw/c)},$$

using the relative F values observed by Lonsdale. The contour map (Fig. 3) shows the result of this computation. If we assume

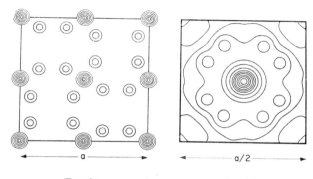

FIG. 2a.   FIG. 2b.

FIG. 2. Potassium dihydrogen phosphate. (a) Electron density projected on (001), J. West.[3] (b) Contour map of the $F^2(hk0)$ series.

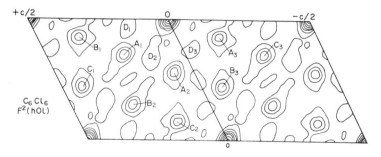

FIG. 3. Hexachlorobenzene. Contour map of the $F^2(h0l)$ series.

that peaks of the type A, B and C are due mainly to Cl–Cl distances and make use of the space group data we are led without further assumptions to a slightly irregular hexagon of chlorines arranged with respect to the axes as shown in Fig. 4.

To explain other principal peaks we are led to an inner hexagon of carbons. The components of the various interatomic distances are given in Table I, as computed from Lonsdale's data (L) and as obtained in the present paper (P). The agreement is, on the whole, very good.

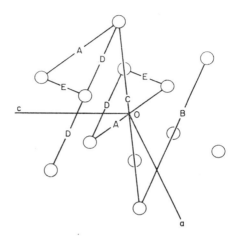

FIG. 4. Interatomic distance diagram for the C_6Cl_6 molecule. The inner ring consists of carbon atoms, the outer of chlorines. The interatomic distances indicated are repeated six times in approximate hexagonal symmetry. The molecule is actually centrosymmetrical. The distances in this diagram give rise to the peaks in Fig. 3 bearing the same letters.

Referring to Fig. 4 we note that the C–C vector (A) is concealed by the similarly labelled Cl–Cl vector. There are also intermolecular Cl–Cl distances very nearly equal to (A) which are largely responsible for the considerable height of peaks of the type (A), (B) and (C) and may also be responsible for their distortion. In Fig. 3 peaks will be seen which correspond to the Cl–C and C–C distances (E) but they are so obviously distorted by neighboring peaks and depressions that it does not seem worth while discussing their positions. In computing the (D) values from

Lonsdale's data an unweighted mean has been taken of the values computed for the three parallel vectors (two Cl–C and one C–C) which contribute to each of these peaks. As this cannot be justified, it seems that the particularly good agreement obtained in this case is fortuitous.

TABLE I.   COMPONENTS OF INTERATOMIC DISTANCES IN HEXACHLOROBENZENE

		A_1	A_2	A_3	B_1	B_2	B_3
u	L	2·25	3·32	1·08	1·17	5·57	4·40
	P	2·38	3·54	1·18	1·18	5·69	4·53
		C_1	C_2	C_3	D_1	D_2	D_3
u	L	4·53	6·89	2·38	0·64	2·66	2·06
	P	4·49	6·65	2·15	0·57	2·66	2·08
		A_1	A_2	A_3	B_1	B_2	B_3
w	L	3·46	1·12	$\overline{2}$·23	5·68	4·57	$\overline{1}$·11
	P	3·69	1·20	$\overline{2}$·49	5·81	4·61	$\overline{1}$·20
		C_1	C_2	C_3	D_1	D_2	D_3
w	L	7·10	2·49	$\overline{4}$·61	2·79	2·16	$\overline{0}$·51
	P	6·80	2·34	$\overline{4}$·57	2·72	2·15	$\overline{0}$·57

6. *Conclusion*

From the preceding examples it is seen that approximate values of the components of interatomic distances can immediately be obtained by plotting "F^2-series" of the type (5), (6) or (7) and in these two cases, with only the space group data in addition, it seems possible to obtain a very good approximate picture of the projection of the structure on the plane which has been investigated.

The values of the interatomic distance components are obviously subject to error from distortion produced by neighboring peaks. It seems, however, that the presence of such distortion will be recognizable from the nature of the series itself. The final values for the interatomic distances must be obtained from a Fourier analysis or a parameter determination of the usual type but the approximate information provided by the F^2-series will eliminate a great many of the possibilities which normally have to be tested by trial and error.

In addition, it seems that in a great many cases where the structure determination cannot be completed, this method will enable some information as to the molecular structure to be obtained from the X-ray data; and it provides a very convenient means of summarizing the information which can be obtained from that data without any extraneous assumptions.

In conclusion the author wishes to thank Professors J. C. Slater and B. E. Warren of this Institute, the former for extending to me the hospitality of the Laboratory, the latter for many valuable discussions.

11 The Diffraction of Slow Neutrons by Crystalline Substances*

W. M. Elsasser

We have examined, from the theoretical point of view, the effect on the elastic scattering of neutrons of interference effects caused by the regular arrangement of atoms in crystals. The wavelength of a neutron which possesses energy kT appropriate to thermal motion is about $1 \cdot 8 \times 10^{-8}$ cm and it is therefore quite close to the interplanar constants of many simple substances.

The selective reflections of the waves in a crystal are given by the well-known Laue–Bragg conditions

$$\kappa \sin \theta = \pi \left| l_1 \vec{b}_1 + l_2 \vec{b}_2 + l_3 \vec{b}_3 \right| = \pi B.$$

Here κ is the wave number, θ the Bragg angle—which is a half of the angle through which the neutron is scattered: the vectors $\vec{b}_i$ are the axes of the reciprocal lattice: l_1, etc., are whole numbers. We shall assume that we are dealing with a polycrystalline powder on which falls a parallel beam of neutrons of different velocities. A single nucleus produces isotropic scattering: the number of neutrons scattered by the powder within an element $d\omega$ of solid angle is given by[1]

$$I = \sum_l \frac{\pi^2 |S_l|^2 M[\bar{\sigma}]}{V \kappa_l^2 \sin^2\theta} f_{(\kappa, l)} \frac{d\omega}{4\pi}.$$

---

*Translated from *Comptes Rendus, Academy of Sciences of Paris*, **202**, 1029–30 (1936).
[1]Laue, M., *Zeits. f. Kristallogr.* **64**, 115 (1926).

The sum includes all the orders l of reflections which make a contribution for a certain angle θ, S_l being the structure factor of each reflection. M is the number of unit cells which comprise the scattering sample, V is the volume of a unit cell, $[\bar{\sigma}]$ is the sum of the scattering cross-sections contained in a cell and, when several isotopes of the same element exist, it is the average sum: finally, $f(\kappa)\,d\kappa$ is the number of neutrons included in the band of wavelengths represented by the interval $d\kappa$.

In order to make an approximate calculation, we have replaced the sum by an integral over the space B (the reciprocal space of the lattice) an integral which extends over the space outside a sphere of radius B_0. Let us assume now that $f(\kappa)$ takes the form of a Maxwell distribution, which is a close approximation to the experimental facts[2]. The formula which follows gives the ratio of the number of scattered neutrons $\mathscr{I}_I$, to the incident number $\mathscr{I}_I{}^0$. Usually one merely measures the activity induced in a rather thin surface layer of a detecting substance and in this case the probability of detection of a neutron will be inversely proportional to its velocity. We have made an effective calculation by introducing first of all a factor $1/\kappa$ in the expression for the scattered intensity, which will be finally denoted by $\mathscr{I}_{II}$. We find as a result of simple calculations

$$\frac{\mathscr{I}_I}{\mathscr{I}_I{}^0} = 8\sqrt{(\pi)}N\bar{\sigma}\left[\int_{B_0\sqrt{\alpha}}^{\infty} e^{-y^2}dy + B_0\sqrt{(\alpha)}e^{\alpha B_0^2}\right]\frac{d\omega}{4\pi}$$

$$\frac{\mathscr{I}_{II}}{\mathscr{I}_{II}{}^0} = N\bar{\sigma}e^{-\alpha B_0^2}d\omega,$$

where
$$\alpha = \frac{h^2}{8\sin^2\theta\, mkT}$$

m being the mass of the neutron, N the number of scattering centres present and $\bar{\sigma}$ the mean cross-section of a centre. For

---

[2]Dunning, J. R. *et al.*, *Phys. Rev.* **48**, 704 (1936).

B_0 we can, at least in the case of simple cubic lattices, take the value corresponding to the lowest selective reflection which occurs. We shall define a critical angle by the relation $\alpha B_0^2 = 1$. This gives $\theta = 26°$ for iron, $\theta = 26°$ for nickel, and $\theta = 25°$ for copper, whereas for the greater part of the other metals the interplanar spacings are larger and the critical angles smaller. For angles smaller than these, then $\mathscr{I}_{II}$ rapidly becomes very small. The experiments begun by Halban and Preiswerk seem indeed to indicate the existence of such an effect.

12   Bragg Reflection of Slow Neutrons*

D. P. MITCHELL and P. N. POWERS

THE peak of the velocity distribution of thermal neutrons[1] indicates a momentum for which the de Broglie wavelength, h/mv, is approximately 1·6Å. If such neutrons suffer Bragg reflection, they will be regularly reflected from a magnesium oxide (MgO) crystal ($2d = 4\cdot0$Å) when the Bragg angle is about $22°$ ($\eta\lambda = 2d\sin\theta_\eta$).

Sixteen well-formed single crystals of MgO, about $8 \times 25 \times 44$ mm, were mounted in a ring with the source and detector placed on the axis for a grazing angle of $22°$, as shown in Fig. 1.

Background, N_{Cd}, of High Speed Neutrons

The detectors were (1), an ionization chamber filled with BF_3 in the first run, and (2), one lined with B_4C in the second and third runs. The sensitivity of both of these chambers[2] extends to neutrons of such high velocity that it was quite impossible to absorb all detectable neutrons emerging in the direction of the chamber. These together with those scattered from the general surroundings account for the number (N_{Cd}) of neutrons counted when the cadmium screening of the chamber is completed by a sheet of Cd across its front.

---

*Physical Review, **50**, 486–7 (1936).

[1]J. R. Dunning, G. B. Pegram, G. A. Fink, D. P. Whitehall and E. Segrè, Phys. Rev. **48**, 704 (1935).

[2]Dana P. Mitchell, Phys. Rev. **49**, 453 (1936).

The entire removal of the crystals made practically no change in the count of these high speed (Cd penetrating) neutrons, and further, the subsequent removal of the Cd from the front of the chamber made no appreciable increase in the count.

Neutrons Scattered by Single Crystals

It thus appears (1), that the crystals do not significantly affect the amount, N_{Cd}, of high speed neutrons counted, and (2), that the slow speed neutrons counted were scattered from the crystals.

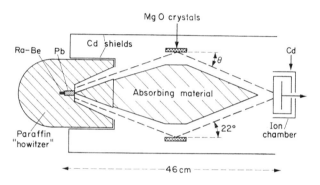

FIG. 1.

When the crystals are in the Bragg positions, the total number N_B of neutrons counted will be the background N_{Cd} plus both those regularly reflected and incoherently scattered by the crystals. The amount of incoherent scattering should be practically independent of crystal orientation, so to observe this without regular reflection the crystals were tilted, alternately clockwise and counter-clockwise, about 25° from the Bragg position. In this case of crossed crystals the total count N_X will be N_{Cd} plus the incoherent scattering. Hence $N_B - N_X$ should be a measure of the number of slow neutrons that are regularly reflected. In the first run, $N_B - N_X$ was eight times that accountable on the basis of statistical fluctuations and in the second run, six times. These results at once indicated the Bragg reflection of slow neutrons. As a check, it seemed necessary to determine by actual test whether or not polycrystalline blocks of about the same size and scattering power would, due to the change in geometric disposition, scatter more slow neutrons in the "Bragg" position than in the crossed position.

TABLE I. OBSERVED NUMBERS

Run	Bragg		Crossed		Background	
	Counts $\times 10^{-3}$	Rate N_B/min.	Counts $\times 10^{-3}$	Rate N_X/min.	Counts $\times 10^{-3}$	Rate N_{Cd}/min.
1st	23	$60.5 \pm .4$	21	$55.6 \pm .4$	3.8	$43.3 \pm .7$
		$N_B - N_X = 4.9 \pm .6$				
2nd	11	$28.8 \pm .3$	8.6	$26.5 \pm .3$	3	$20.9 \pm .4$
		$N_B - N_X = 2.3 \pm .4$				
3rd	12	$37.6 \pm .3$	12	$37.7 \pm .3$	6	$28.0 \pm .4$
		$N_B - N_X = .1 \pm .4$				

TABLE II. RELATIVE NUMBERS

Run	$\dfrac{N_B - N_X}{N_B - N_{Cd}}$	$\dfrac{N_X}{N_{Cd}}$	$\dfrac{N_B}{N_{Cd}}$
	With MgO Crystals		
1st	0.40 ± 0.06	1.28 ± 0.02	1.40 ± 0.02
2nd	$.41 \pm .09$	$1.27 \pm .03$	$1.38 \pm .03$
	With Al Blocks		
3rd	0.01 ± 0.04	1.34 ± 0.02	1.34 ± 0.02
	corrected	1.22	1.22

Scattering by Polycrystalline Blocks

In this, the third run, aluminium metal blocks of rectangular size and thickness equal to the rectangular boundary of the somewhat irregular single crystals, used in the first two runs, were mounted in place of the crystals. Aluminium was used since it has approximately the same effective scattering power per unit volume as MgO. The result

$$N_B - N_X = 0 \cdot 1 \pm 0 \cdot 4$$

indicates that the change in geometry could not have changed the incoherent scattering much more than the statistical fluctuations.

Summary of Results

The greatest statistical precision was obtained for N_B and N_X, and hence these two values in Table I are the best evidence for the Bragg type of reflection of neutrons. It is also interesting to note the relative amount of reflection and scattering as shown in Table II.

The correction shown takes account of the fact that the total volume of the aluminium blocks used in the third run was $1 \cdot 63$ times that of the crystals (due to the irregular outline of the crystals, as mentioned above).

On the basis of this evidence it seems reasonable to conclude that we have in these experiments observed the reflection of slow neutrons in accord with the Bragg relation between the de Broglie wavelength of these neutrons and the grating space of these crystals.

It should be noted that the experimental arrangement (see Fig. 1) permits a sufficiently large angular divergence so that the Bragg conditions are satisfied for a large portion of the velocity range in the Maxwellian distribution[1] of the thermal neutrons.

Grateful acknowledgment is made to Mr. Raymond Ridgeway for supplying us with the unusually large single crystals of MgO used in this work.

Pupin Physics Laboratories, Columbia University,
August 17, 1936.

13 Diffraction of Neutrons by a Single Crystal*

W. H. Zinn†

The intensity of neutrons emitted by the nuclear chain reactor has been shown to be sufficiently great to permit the observation of the Bragg reflection of neutrons by a single crystal. A single crystal monochromator has been constructed and used for the investigation of resonances. The well-known resonance in Cd has been measured and the Breit-Wigner constants determined.

Introduction

A number of experiments designed to demonstrate the diffraction of slow neutrons by crystals have been made in the years following the discovery of the neutron. Early methods of treating the interaction between the neutron and the nucleus, developed by Fermi[1] and extended by Wick,[2] anticipated difference in scattering in crystalline and disordered substances. Experimental verification was attempted by Preiswerk and von Halban[3] who measured the temperature variation in the distribution of neutrons transmitted by a polycrystalline medium, and Mitchell and Powers,[4] as well as Preiswerk,[5] presented evidence for a coherent component of a neutron beam scattered from an array of crystals.

*Physical Review, **71**, 752–7 (1947).

†Argonne Laboratory, University of Chicago, Chicago, Illinois. Received February 26, 1947.

[1]E. Fermi, *Ricerca Scient.* **7**, pt. 2, 13 (1936).

[2]Wick, *Physik. Zeits.* **38**, 403 (1937).

[3]Von Halban and Preiswerk, *Comptes rendus* **203**, 73 (1936).

[4]Mitchell and Powers, *Phys. Rev.* **50**, 486 (1936).

[5]Preiswerk, *Helv. Phys. Acta* **10**, 400 (1937).

More recently, further observations of neutron diffraction have been made in the measurements of slow neutron scattering by Whittaker and Beyer[6] and in the work of Anderson, Fermi, and L. Marshall[7] and of Rainwater and Havens[8] who observed variations with the energy of the neutron beam in the scattering of crystalline substances. The high neutron flux generated by the chain reacting pile has permitted, as reported here, a series of

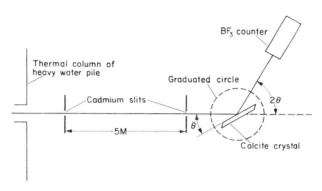

FIG. 1. First experimental arrangement.

experiments which show the existence of an intense diffracted beam of neutrons from a single crystal, and which demonstrate that this beam is of sufficient intensity for the study of the energy dependence of neutron reactions by methods similar to those of X-ray spectrometry. The work of this paper has been reported briefly[9] and similar results have been given by Borst *et al.*[10]

[6]Whittaker and Beyer, *Phys. Rev.* **55**, 1101 (1939); Beyer and Whittaker, *ibid.* **57**, 976 (1937).

[7]Anderson, Fermi, and L. Marshall, *Phys. Rev.* **70**, 102 (1946).

[8]Havens and Rainwater, *Phys. Rev.* **70**, 154 (1946); Rainwater and Havens, *ibid.* **70**, 136 (1946).

[9]W. H. Zinn, *Phys. Rev.* **70**, 102A (1946).

[10]L. B. Borst, A. J. Ulrich, C. L. Osborne, and B. Hasbrouck, *Phys. Rev.* **70**, 108A (1946); *ibid*, **70**, 557 (1946).

Description of Apparatus

Neutrons diffusing from the main part of the chain reacting pile are permitted to enter a large block of graphite which acts as a slowing down column, bringing the average neutron energy to the equilibrium value for graphite at room temperature. Since the de Broglie wavelength of these neutrons is of the same order of magnitude as that of X-rays, it was possible to attempt the

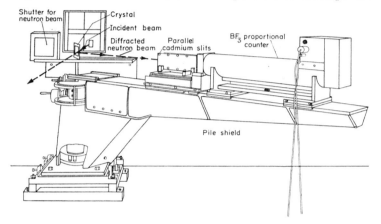

FIG. 2. The crystal spectrometer.

diffraction experiments with the customary crystals, using a spectrometer similar to that used with X-rays. As is shown schematically in Fig. 1, a collimated beam of neutrons was obtained by introducing two thick cadmium slits, 5 meters apart, in the neutron beam from the graphite thermal column of the heavy water pile. Cadmium slits sufficed here since the neutron radiation from a thermal column is singularly free from fast neutrons. The collimated beam fell upon the (1, 0, 0) planes of a large single calcite crystal, and the diffracted beam was detected by means of a BF_3 proportional counter filled, for greater detection efficiency, with gas enriched in the B^{10} isotope. Only early experiments were performed at the thermal column;

subsequently the apparatus was set up at a side hole of the reactor where the high energy neutron flux is much greater than at the thermal column.

The final form of the spectrometer (Fig. 2) is identical in principle with the usual X-ray instrument; however, in order to support the heavier collimators and detectors which are necessary for neutron diffraction experiments, the unit is made very massive. To improve the resolution the arm is long, having a length of 80 inches and a width of 8 inches, A 6-inch diameter crystal table is provided for mounting the crystals. The two graduated circles indicating the position of the arm and crystal table are 12 inches in diameter and are calibrated in degrees, while greater precision is attainable from readings on a cylindrical scale calibrated in quarter minutes and placed at the drive wheel. The table and arm are driven by means of large knurled wheels through a train of gears, and in use can be adjusted independently or coupled by a friction clutch. The table and arm are driven in the two to one angular ratio necessary for reflection experiments.

In order to measure with greatest efficiency the flux of diffracted neutrons, the beam was passed axially through the proportional counter. In this manner it was possible to detect nearly 100 percent of all the neutrons of the low energies. The counter was 5 cm in diameter, 60 cm long, and filled to a pressure of 40 cm of Hg with boron trifluoride. This gas was enriched about five times in the B^{10} isotope and thus transmitted but 5 percent of neutrons of kT energy. The required degree of collimation was obtained by inserting into the shield of the reactor a steel block containing a long channel. This channel was approximately $\frac{1}{2}''$ wide, $1''$ tall, and $8'$ long. The channel was directed at a part of the reactor containing heavy water and uranium. The beam collimated by this means contains neutrons ranging in energy from thermal to those of the fission spectrum itself. The intensity obtainable is a sharp function of the degree of collimation desired and the spectral region to be investigated. At the maximum of the thermal spectrum, with quite excellent resolution, counting rates of 5000 counts per minute can be obtained. At the suggestion of

W. H. Zachariasen, roughening the crystal surfaces was tried and gives an improvement in intensity of about a factor of two. At energies in the so-called resonance region considerably smaller

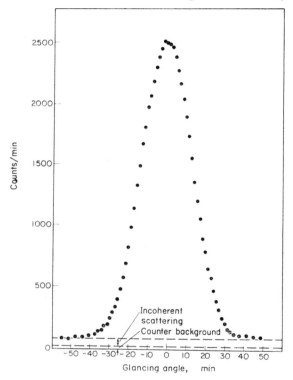

FIG. 3. Rocking curve (1, 0, 0) LiF crystal.

counting rates are obtained. A detector more sensitive to neutrons in this region is very much to be desired.

Results

A typical rocking curve is shown in Fig. 3. No attempt has been made to correct these data for resolution or detector sensitivity. The detector, boron, follows a $1/v$ law for absorption, and as a result the sensitivity of the counter is not uniform over the entire range of velocities. Therefore, the rocking curve is not exactly symmetrical, showing relatively higher values on the low energy side of the diffraction peak.

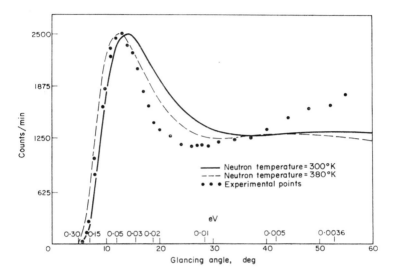

FIG. 4. Spectrum of thermal column radiation, calcite crystal. Solid line is the theoretical curve calculated by Goldberger and Seitz.

Sufficient intensity of diffracted neutrons existed throughout the range of energies emitted from the thermal column to make measurement of its neutron spectrum possible by this method. Intensities at low energies, since they are contaminated by higher order components and are more efficiently detected, show considerably higher values than the real distribution in this region. Figure 4 shows the distribution of neutrons from the thermal column as obtained with a calcite crystal. Figure 5 shows the spectrum of neutrons directly from the reactor. These neutrons have not diffused through any considerable column of moderator material. Again no correction to the data has been made for detector sensitivity or for resolution. Neither of the corrections is sizeable in the low energy region covered in these curves. Conversion from glancing angle to energy was made by the relation

$$ev = (2 \cdot 22 \times 10^{-3})/\sin^2\theta,$$

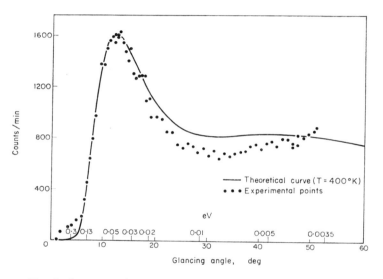

FIG. 5. Spectrum of reactor radiation, calcite crystal. Solid line is the theoretical curve calculated by Goldberger and Seitz.

where θ is the glancing angle. Measurements were made in the energy interval between 0·004 ev and 0·30 ev, which is in the range of glancing angles between 50° and 2°. Both spectra show a strong, approximately Maxwellian component, distorted by high order contributions on the long wavelength side of the maximum. The spectrum of neutrons from the wall of the reactor, having passed through a smaller amount of moderator, has an additional component at small glancing angles representing neutrons not yet in equilibrium with the moderator. Attempts to fit the experimental data theoretically have been made by Goldberger and Seitz[11] for various moderator temperatures and are included on the figures.

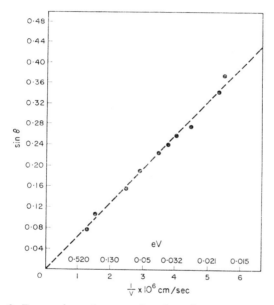

FIG. 6. Boron absorption as a function of neutron energy. The dotted curve is a plot of $(h/mv) = 2d \sin\theta$. The points are experimentally determined values of $1/v$ and $\sin\theta$.

To demonstrate that the diffracted beam consists primarily of neutrons in a very small energy band, a measurement of the transmission of a Pyrex plate was made over the range 0·018 to 0·5 ev. This plate, previously standardized by means of a mechanical velocity selector, was shown to have a total cross section,

[11]Goldberger and Seitz, *Phys. Rev.* **70,** 116 (1946). A complete discussion of their method will be published shortly.

because of its boron content, very close to $1/v$ in the region investigated. With the calcite crystal, measurements were made of the transmission of the plate from which effective neutron velocities were obtained by comparison with the measurements made on the mechanical velocity selector. Reciprocals of these velocities as a function of the sine of the glancing angle are plotted as points in Fig. 6. The dotted curve gives the reciprocal velocity

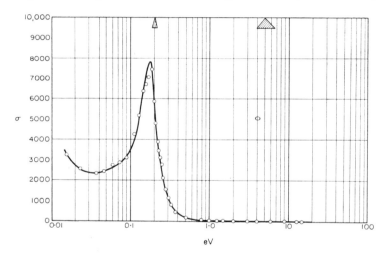

FIG. 7. Total cross section of cadmium. Solid curve is Breit-Wigner fit to experimental points. σ in units of 10^{-24} cm² per atom.

as calculated from the glancing angle and the crystal spacing. Similar measurements show that the straight line behavior shown extends to 15 volts which was the upper limit of the instrument at the time these measurements were made.

The total cross section of cadmium has been measured by the method of the modulated cyclotron beam[12] and a single strong level observed. The results obtained for cadmium with the crystal monochromator are shown in Fig. 7. The constants of the

[12]C. P. Baker and R. F. Bacher, *Phys. Rev.* **59**, 332 (1941).

resonance are similar to those of Rainwater and Havens[8] but show a greater maximum than those of Baker and Bacher. The measurements of Fig. 7 extend to 15 electron volts, and no further resonance structure is found. A series of six foils of cadmium of thickness varying from 0.000142×10^{24} to 0.115×10^{24} atoms per square centimeter were required for the measurement. Wherever possible the range of acceptable transmissions was held between 0.50 and 0.75 to reduce effects of high energy beam components. The experimental results are best fitted with a curve of the Breit-Wigner function having the constants: $E_0 = 0.180$ ev, $\sigma_0 = 7800 \times 10^{-24}$ cm^2/atom, $\Gamma = 0.122$ ev, and $\sigma_s = 6.0 \times 10^{-24}$ cm^2/atom.

Acknowledgment is hereby made for the invaluable advice and skill in construction of the instrument contributed by T. J. O'Donnell, D. DiCostanzo, and J. Getzholtz of the laboratory workshop.

The initial results reported here were obtained in July–September, 1944. This work was carried out under the auspices of the Manhattan District.

14 Interference Phenomena of Slow Neutrons[*]

E. Fermi and L. Marshall[†]

Various experiments involving interference of slow neutrons have been performed in order to determine the phase of the scattered neutron wave with respect to the primary neutron wave. Theoretically this phase change is very close to either 0° or 180°. The experiments show that with few exceptions the latter is the case. The evidence is based on the following types of measurements: (a) measurement of the intensities of Bragg reflection of various orders of many crystals, and comparison with the theoretical values of the form factor; (b) total scattering cross section of gas molecules for wavelengths long compared with the molecular dimensions; and (c) determination of the limiting angles for total reflection of neutrons on various mirrors. The elements Ba, Be, C, Ca, Cu, F, Fe, Mg, N, Ni, O, Pb, S, and Zn were found to scatter neutrons with 180° phase difference; Li and probably Mn scatter with zero phase difference. The five elements I, Br, Cl, K, and Na behave alike and the phase with which they scatter is tentatively identified as 180°. Coherent scattering cross sections have been determined for several elements.

Introduction

The scattering processes of slow neutrons are greatly complicated by interference phenomena due to the fact that the de Broglie wavelength is comparable with interatomic distances. The general pattern of interference phenomena of slow neutrons is similar to that of X-rays, since both the wavelength and the scattering cross section of X-rays are comparable to those of neutrons. On the other hand there are considerable differences due to several factors. Among them is the fact that the scattering

*Physical Review, **71,** 666–77 (1947).

†Argonne National Laboratory and University of Chicago, Chicago, Illinois. Received February 7, 1947.

of X-rays varies regularly with atomic number while that of neutrons is a rather erratic property. Furthermore, in the case of neutrons the phase difference between scattered and incident wave may be either 0° or 180° as will be discussed in Section 1. For X-rays instead, it is always 180° because X-ray energies are larger than most electronic resonance energies. Also, the absorption properties of neutrons differ markedly from those of X-rays.

The main purpose of this work was the investigation of various interference phenomena in order to determine the phase change of the scattered neutron wave for a large number of elements. Section 1 contains a summary of the theoretical background of this work. Section 2 describes the measurements of the intensities of Bragg reflections of various orders and their interpretation. Section 3 is a discussion of some experiments on filtered neutrons. In Section 4, experiments on scattering of neutrons by gas molecules are presented. Section 5 describes measurements of the limiting angle for total reflection of neutrons. The general conclusions are discussed in Section 6.

1. *Theoretical Considerations*

When a slow neutron is scattered by a nucleus, its de Broglie wave at some distance from the scattering nucleus may be written as the sum of a term $\exp(ikx)$ representing the primary wave and a term $-a[\exp(ikr)]/r$ representing the scattered wave. This is true for the case of slow neutrons because with very good approximation the scattering is spherically symmetrical. The coefficient of the latter term has been written with the minus sign for reasons of convenience that will be apparent later. If the constant a is positive, there is a phase change of 180° between scattered and incident wave, and if a is negative, the phase change is 0°. One proves in an elementary way that the scattering cross section is related to the constant a by the equation

$$\sigma = 4\pi|a|^2 \tag{1}$$

and also that the constant a is with very good approximation for slow neutrons a real number. Its imaginary part is very small and can usually be neglected except in case of extremely high absorption. The quantity a which has the dimensions of a length and the order of magnitude of 10^{-12} cm shall be referred to as the "scattering length". The main purpose of this paper is the experimental determination of the scattering length and in particular of its sign.

For elements consisting of one isotope only and without nuclear spin, the magnitude of the scattering length can be immediately obtained from (1), so that only its sign needs to be determined. Even in this simple case a small correction must be applied depending on whether the atom is free or bound. Because of the change in the reduced mass, the cross section of a free atom differs from that of the same atom bound in a crystal, by a factor $[(A+1)/A]^2$, where A is the atomic weight.[1]

Since the scattering length is proportional to the square root of the cross section, the correction factor will be $(A+1)/A$. It follows that

$$a = [(A+1)/A]a_f \qquad (2)$$

where a indicates the scattering length of the atom bound in a crystal and a_f the scattering length for the free atom. For elements that are mixtures of several isotopes with probabilities of occurrence $p_1, p_2, \ldots, p_n$, and scattering lengths $a_1, a_2, \ldots, a_n$ the scattering length is the average

$$a = p_1a_1 + p_2a_2 + \ldots + p_na_n. \qquad (3)$$

This is the magnitude that determines the interference properties of the element. The scattering cross section is no longer given by (1) but by

$$\sigma = 4\pi\{p_1a_1{}^2 + p_2a_2{}^2 + \ldots + p_na_n{}^2\}. \qquad (4)$$

Actually in this case the relationship between a and σ can be expressed by the inequality

$$|a| \leqq (\sigma/4\pi)^{\tfrac{1}{2}}. \qquad (5)$$

[1] For example see H. A. Bethe, *Rev. Mod. Phys.* **9**, 71 (1937).

The knowledge of the scattering cross section is insufficient, therefore, to determine even the magnitude of the scattering length.

A similar situation obtains if the nucleus has a spin I different from zero. Here the spin vector of the scattering nucleus is oriented either parallel (total spin $I+\frac{1}{2}$), or antiparallel (total spin $I-\frac{1}{2}$) to the spin vector of the neutron. Consequently the scattering length may take either of two values, $a_{I+\frac{1}{2}}$ or $a_{I-\frac{1}{2}}$. The effective scattering length is the average of these two scattering lengths, each being weighted by the probability of occurrence of the corresponding spin orientation, namely

$$a = \frac{I}{2I+1}a_{I-\frac{1}{2}} + \frac{I+1}{2I+1}a_{I+\frac{1}{2}}. \tag{6}$$

The scattering length will obey (5), the equal sign corresponding to the case that $a_{I+\frac{1}{2}}$ is equal to $a_{I-\frac{1}{2}}$ both in magnitude and sign.

A simple geometrical interpretation of the scattering length can be obtained if the interaction between the neutron and the nucleus is represented by a potential well. Let the eigen-function of the neutron for an s state of energy zero be $\psi(r)$. In Fig. 1, $r\psi(r)$ is plotted *versus* r. The curve becomes a straight line for r greater than the radius r_0 of the potential well. This straight line can be extended to intersect the r axis at the point P. One can prove that the scattering length a is the abscissa of the point P. Figure 1a corresponds to a case in which a is positive, and Fig. 1b to a case in which a is negative. By inspecting the two figures it is apparent that the probability is much higher for a positive than for a negative, especially for heavy elements. This remark on the relative probability for a to be positive or negative is due to E. Teller and V. Weisskopf.

A simple formula can be obtained in the case that the scattering is due to the effect of a single Breit-Wigner resonance level. It is, then

$$a = -\lambda_R \Gamma_n / [W - R + i\Gamma] \tag{7}$$

where W is the energy of the neutron, R is the resonance energy,

Γ_n and Γ are the neutron and total half width at half maximum, and λ_R is the de Broglie wavelength at energy R, divided by 2π. Since in most cases Γ is small compared with $(W - R)$, it follows that a will be positive or negative depending on whether the energy of the neutron lies below or above the resonance energy. This picture is a great oversimplification since there are no cases

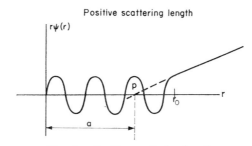

FIG. 1a. Positive scattering length.

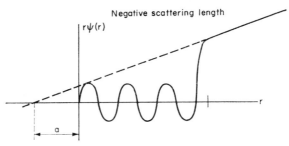

FIG. 1b. Negative scattering length.

in which the Breit-Wigner scattering of a single level is the dominant phenomenon, and complicated situations arise due to interference of Breit-Wigner and potential scattering.

2. Intensities of Various Orders of Bragg Reflection

One can compare the changes in phase in the neutron scattering of different elements by measuring the intensity of the Bragg scattering for various orders and various crystalline planes of crystals containing at least two elements.

The simplest case is that of a crystal in which the planes are equidistant and consist alternately of two kinds of atoms, as for example, the 1, 1, 1 planes of NaCl. These planes are equidistant and consist alternately of sodium and of chlorine. In the first order Bragg reflection, the optical path for reflection from sodium planes differs from the path for reflection from chlorine planes by $\lambda/2$. Consequently if sodium and chlorine nuclei cause the same change in phase of the scattered neutron wave, their contributions will subtract and the order will have low intensity. If they scatter with opposite phase change their contributions will add and the order will have high intensity.

The situation for the second order is reversed. Here the difference in optical path for reflection from the two kinds of planes is λ. Consequently if sodium and chlorine scatter with the same change in phase the order will have high intensity, and if with opposite change in phase, low intensity. Similarly it follows that if the two kinds of nuclei scatter with the same change in phase the odd orders will be weak and the even orders will be strong. Conversely, if they scatter with opposite change in phase, the odd orders will be strong and the even orders weak.

Analogously to X-ray scattering, there is superimposed upon this effect a continuous decrease in intensity from order to order due to geometrical factors, to thermal agitation, and to imperfections of the crystal.

For more complicated cases the intensity of the various orders is determined as for X-rays by the form factor

$$F = \left| \sum {}_j a_j \exp(2\pi i n \delta_j/d) \right| \tag{8}$$

where a_j is the scattering length, d is the spacing of the lattice planes, n is the order of the Bragg reflection, and δ_j is the perpendicular distance from the jth atom to the plane of reflection. Naturally the form factor is strongly dependent on the relative signs of the scattering lengths of the various atoms of the crystal. An analysis of the intensities for various orders and various planes of the crystal will often allow the determination of the relative signs of the scattering lengths of the elements involved. In several cases it is possible to determine also the ratios of the values of the scattering lengths.

A comparison of intensities for Bragg reflections of various orders of many crystals have been made using monochromatic neutrons. The experimental arrangement is schematically represented in Fig. 2. A beam of non-monochromatic thermal neutrons emerges from a $4'' \times 4''$ hole containing a long collimator $\frac{1}{2}''$ wide by $1\frac{1}{4}''$ high in the thermal column of the Argonne heavy water pile. This beam falls on the 1, 0, 0 plane of a large CaF_2

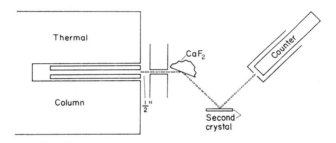

FIG. 2. Apparatus for measuring intensity of Bragg orders.

crystal at a glancing angle of about 16°. Neutrons of energies satisfying the Bragg condition ($\lambda \sim 1.50$ angstroms for the first order) are reflected to form a beam that is used for investigating various crystal specimens. The beam contains a main component of neutrons of wavelength 1.50Å plus a small fraction of neutrons of energy four times as large due to the second order reflection on the CaF_2 crystal. The second order component is fairly unimportant because there is only a small number of neutrons of this relatively high energy in the Maxwell distribution. We can, therefore, consider the beam reflected on the calcite to be approximately monochromatic. This monochromatic beam of neutrons falls on some plane of the second crystal, which is mounted on a rotating table. The neutrons undergo a second Bragg reflection, and are finally detected by an enriched BF_3 proportional counter. The counter is supported by an arm which rotates about the axis of the crystal table. Additional Cd slits not shown are used for more precise collimation of the neutron beam.

A typical measurement was carried out as follows. Both counter and second crystal were set approximately at angles corresponding to a reflection of a given order. By rocking each separately the setting of maximum intensity was found. Also the inclination of the crystal was adjusted for maximum intensity. From the intensity measured in these conditions was subtracted a background, measured by turning the crystal a small angle off the Bragg position. In the majority of cases the background was a small fraction of the total count.

The results of these measurements are summarized in Table I. Columns 1 and 2 indicate the crystal and the plane. Column 3 is the form factor. Here the chemical symbol of the element has been used to represent its scattering length. The fourth column gives the absolute value of the form factor, using the values of the scattering length given in Table VII. The fifth column gives the measured intensities of the various orders in counts per minute. In the sixth column is the ratio of column 5 to column 4. If the values chosen for the scattering lengths are approximately correct, one expects that the values in column 6, corresponding to any given plane, should show a regular decrease in intensity with increasing order, since dividing the intensity by the form factor should correct it for the irregular change of intensity from order to order. Therefore only the regular decrease should remain.

There may be some doubt whether it is more appropriate to divide the intensity by the form factor or by its square. Theoretically one would expect that, for ideally perfect crystals, the form factor should be used, and for ideally imperfect crystals, its square. Actually, we have found that one obtains a much better fit by using the form factor.

For the simplest crystals containing two elements one can see immediately by inspecting the intensity data of column 5 whether the scattering lengths of the two elements have equal or opposite sign. For instance CaF_2 (100), NaCl (111), and PbS (111) show clearly superimposed on the general decrease of intensity with increasing order, an alternation of intensity with strong even orders and weak odd orders. Since, in these crystals, the planes consist alternately of the two kinds of atoms, we conclude that the scattering lengths of each pair have the same sign. The opposite is the case for LiF (111) where the even orders have intensity so low that we could not measure them, and the odd orders have a normal intensity. Therefore Li and F have scattering lengths of opposite sign. In some of the more complicated cases the analysis is also quite straight forward as can be seen by comparison of the observed intensities with the formulae for the form factors in column 3. In a few cases the interpretation is not unique.

The following conclusions as to the signs of the scattering lengths can be drawn. The components of the following pairs have scattering lengths of the same sign: Na, Cl; Pb, S; Ca, F; Fe, S; Fe, O; K, Br; Mg, O; K, Cl; K, I. The components of the pair, Li, F have scattering lengths of opposite sign. The measurements of MnS_2 do not allow a unique interpretation although the probable conclusion is that Mn and S also have scattering lengths of opposite sign.

Using the evidence to be presented in Sections 4 and 5 that carbon and oxygen have scattering lengths with the same sign, one can conclude from the measurements on calcite that Ca and O also have scattering lengths of the same sign.

From the measurements on FeS_2 and Fe_3O_4, one can conclude that S and O have the same sign; and from data on $BaSO_4$, one is led to assign also the same sign to Ba. The measurements on $NaNO_3$ indicate that N and O have the same sign. (See confirmatory evidence in Section 4.) The measurements of the (111) plane of $NaNO_3$ indicate although not quite conclusively that Na has the same sign as the group NO_3, and therefore as N and O.

From these data it follows that C, O, Fe, Mg, Ba, Ca, S, F, Pb, and N all have the same sign. Na, K, Cl, Br, I all have like sign. The partial evidence just mentioned, according to which Na has the same sign as O, indicates that all this latter group

TABLE I. INTENSITY OF BRAGG ORDERS

Crystal	Plane	Order	Form Factor		Intensity	Ratio
CaF$_2$	(100)	1	Ca$-$2 F	·41	16300	39800
		2	Ca$+$2 F	1·99	20300	10200
		3	Ca$-$2 F	·41	1287	3140
NaCl	(111)	1	Na$-$Cl	·57	2376	3990
		2	Na$+$Cl	1·69	2750	1630
PbS	(111)	1	Pb$-$S	·20	7280	36400
		2	Pb$+$S	·76	10700	14100
		3	Pb$-$S	·20	808	4040
		4	Pb$+$S	·76	750	986
PbS	(100)	1	Pb$+$S	·76	19650	25800
		2	Pb$+$S	·76	11420	15000
		3	Pb$+$S	·76	2249	2960
FeS$_2$	(100)	1	Fe$+$ ·62 S	·99	6893	6960
		2	Fe$-$1·6 S	·37	1354	3660
		3	Fe$-$1·6 S	·37	726	1960
MnS$_2$	(111)	1	Mn$-$1·06 S	·74	7930	10700
		2	Mn$+$ ·06 S	·42	2560	6100
		3	Mn$+$ ·06 S	·42	670	1600
Fe$_3$O$_4$	(111)	1	2·34 Fe$+$ 1·20 O	2·65	21300	8040
(magnetite)		2	16 Fe$-$31·9 O	6·34	19970	3150
		3	13·7 Fe$-$ 3·55 O	9·06	12570	1390
		4	8 Fe$+$31·50 O	25·78	16700	647
		5	13·7 Fe$+$ 5·94 O	14·85	5120	344
CaCO$_3$	(111)	1	2 C$+$6 O$-$2 Ca	3·42	9500	2780
(calcite)		2	2 C$+$6 O$+$2 Ca	6·58	8520	1290
		3	2 C$+$6 O$-$2 Ca	3·42	1840	538
CaCO$_3$	(211)	1	2 Ca$+$2 C$+$1·76 O	3·99	10900	2730
cleavage		2	2 Ca$+$2 C$-$2·0 O	1·70	3360	1980
		3	2 Ca$+$2 C$+$2·7 O	4·57	2400	528

should be included with the former. That this conclusion is correct is strongly supported by evidence to be presented in

K

TABLE I.—*Continued*

Crystal	Plane	Order	Form Factor		Intensity	Ratio
$BaSO_4$	(100)	1	4 Ba+4 S+ ·42 O	4·54	3606	794
		2	4 Ba+4 S+14·6 O	13·19	4837	366
		3	4 Ba+4 S+ 3·1 O	6·17	756	123
$BaSO_4$	(011)	1	−3·4 Ba+3·8 S+4·7 O	1·24	1018	821
		2	1·75 Ba+3·1 S−9·6 O	3·61	1632	452
		3	·42 Ba+2·0 S−7·9 O	3·93	651	166
		4	−2·46 Ba+ ·75 S−2·2 O	3·07	36	12
$BaSO_4$	(101)	1	·87 Ba−1·25 S− 1·47 O	·56	737	1320
		2	1·13 Ba−2·02 S− 4·56 O	2·46	1946	791
		3	3·77 Ba+ ·98 S+ 2·56 O	4·81	1477	307
		4	·27 Ba− ·10 S− 2·12 O	1·47	—	—
		5	1·04 Ba+2·12 S−10·4 O	4·93	587	119
MgO	(100)	1	Mg+O	1·17	10352	8800
		2	Mg+O	1·17	6258	5350
MgO	(111)	1	Mg−O	·05	764	15300
		2	Mg+O	1·17	14175	12100
		3	Mg−O	·05	132	2640
LiF	(111)	1	Li−F	1·19	10080	8470
		2	Li+F	·01	~0	—
		3	Li−F	1·19	300	252
KBr	(111)	1	Br−K	·21	1545	7360
		2	Br+K	·91	2853	3140
KCl	(111)	1	Cl−K	·78	2346	3010
		2	Cl+K	1·48	5620	3800
		3	Cl−K	·78	334	428
		4	Cl+K	1·48	160	108
$NaNO_3$	(211)	1	Na+N+ ·88 O	1·96	29000	14800
		2	Na+N− ·98 O	·83	4171	5000
		3	Na+N+1·34 O	2·25	4148	1800
$NaNO_3$	(111)	1	NO_3−Na	2·14	29400	13700
		2	NO_3+Na	3·26	16790	5150
		3	NO_3−Na	2·14	2243	1050
KI	(111)	1	K−I	·01	85	—
		2	K+I	·71	348	500
		3	K−I	·01	24	—

Section 6. Li and perhaps Mn have scattering lengths of the other sign. They represent together with hydrogen (see Section 6) the only exceptions found so far to the behavior of the majority of elements.

This type of data allows only a comparison of the signs of scattering lengths of different elements but not their absolute determination. The absolute determination of sign of scattering length will be discussed in Section 5. The over-all conclusions from Section 2 will be presented in Table VII. There, an attempt has been made to give the actual value of the scattering lengths for several elements. In calculating these values the attempt was made to obtain the best possible agreement of the observed intensities given in Table I, with the form factor after proper account was taken of the natural decrease of intensity with increasing order.

In X-ray analysis of crystals the Debye-Scherrer method of powder photography is extensively used. For neutrons, however, the intensity of the Debye-Scherrer maxima is rather small, and in most cases the method will be impractical. For example, in the case of microcrystalline graphite, we were able to detect only the maximum corresponding to the first reflection on the 001 plane.

A second possibility to study microcrystalline substances is to measure the total cross section of the substance for neutrons of wavelength close to $2d$ where d is the longest lattice spacing. An analysis of this case for substances containing two elements has been made by Fermi and Sachs, and indicates that the total cross section in this region depends strongly on whether the scattering lengths of the two elements have equal or opposite sign. The practicability of this method is somewhat limited by the fact that the theory applies only when the microcrystalline grain of the substance is exceedingly fine.

3. Spectrum of Filtered Neutrons

In the study of interference in gas molecules, somewhat simpler results are obtained with neutrons of energy far below thermal energies and which have a wavelength long compared with inter-atomic distances.   Such neutrons are present in very small percentage in the Maxwell distribution at room temperature. Consequently, one cannot isolate such low energy neutrons by Bragg reflection because in the reflected beam there is a contamination by high orders which is many times more intensive

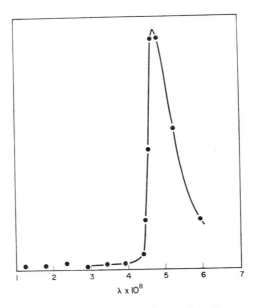

$\lambda \times 10^8$

FIG. 3. Spectrum filtered through BeO.

than the first order. A simple way to avoid this difficulty, without use of extremely low temperature moderators, is to remove the high energy part of the Maxwell distribution by passing the beam of neutrons through a filter of a microcrystalline material of low absorption,[2] such as graphite or BeO. This filtering

[2] Anderson, Fermi, and Marshall, *Phys. Rev.* **70**, 815 (1946).

action of microcrystalline substances is due to the fact that neutrons of wavelength longer than $2d$, where d is the longest lattice spacing of the crystal, cannot satisfy the Bragg reflection condition for any of the many crystallites that they meet while crossing the filter. Consequently, they are scattered out of the beam only to a minimal extent. The production of very low energy neutrons by the filtering process has been previously reported[2] for the case of graphite, where the limiting wavelength is 6·7Å. In the present experiments BeO filters have been used, for which the limiting wavelength is 4·4Å. In order to get experimental proof of the correctness of the interpretation of the filtering process, we have investigated with the crystal spectrometer the spectrum of the neutrons filtered by a prism 40 cm long and containing 100 g/cm². The side dimensions of the prism were about 10×10 cm. The analysis was carried out with a crystal of celestite in the 001 plane. The results are plotted in Fig. 3, where the wavelengths are plotted on the abscissae, and the observed intensities are the ordinates. The figure shows clearly that the spectrum of the filtered neutrons has negligible intensity for short wavelength, rises abruptly to a maximum value at wavelength of approximately 4·5A, and after that, decreases gradually with increasing wavelength.

The decrease of intensity on the long wavelength side is somewhat more rapid than corresponds to the Maxwell distribution. This probably is due to the fact that the neutrons emerging from the graphite column are not fully slowed down to the lowest energies of the Maxwell distribution by the collisions in graphite.

The fact that microcrystalline BeO and Be metal are good filters for neutrons is an indication that the scattering lengths of Be for the two spin orientations of the neutron do not differ very considerably. Indeed if this were the case one would expect a strong incoherent scattering that should not vanish even for wavelengths longer than $2d$. For filtered neutrons we found a residual cross section for BeO of 0.7×10^{-24} cm². If all this residual cross section were due to the difference in the scattering lengths a_1 and a_2 for the two spin orientations, it could be calculated that a_1 and a_2 certainly have the same sign and that their ratio is about 2. Probably the actual values are appreciably closer, because part of the observed residual cross section is certainly due to crystal imperfections and to thermal agitation.

4. *Scattering of Neutrons by Gas Molecules*

Some conclusions on the scattering of neutrons can be drawn from the study of the total cross section of gas molecules. For neutrons of wavelength short compared with interatomic distances the molecular scattering cross section is the sum of the individual scattering cross sections of the atoms in the molecule. When neutrons of longer wavelength are used, a number of complications arise, partly due to interference effects, and partly to the fact that the scattering now is due to atoms that no longer can be considered free.

The theoretical calculation of the cross section has been carried out by Teller and Schwinger[3] for the case of the H_2 molecule for low rotational states. For heavier molecules, such as N_2 or O_2, even when low temperatures and slow neutrons are used, a fairly large number of rotational states is always excited so that a detailed calculation of the contributions of all rotational states becomes impractical. Simple results can be calculated by neglecting the neutron mass in comparison with the mass of the atoms in the molecule. In this approximation one obtains results identical with those of the classical theory of interference. The approximation is quite good in the case of X-rays, where the particle scattered is a photon of very small effective mass, and has been used currently by Debye and his co-workers. For scattering of neutrons from molecules such as N_2, O_2, etc., the approximation is not nearly as good. The corrections, however, have not been calculated, and the results will be compared with those of the approximate theory except in the case of H_2 where the theory of Teller and Schwinger has been used.

---

[3]Teller and Schwinger, *Phys. Rev.* **52**, 286 (1937).

The scattering cross sections of several molecules have been determined using very low energy neutrons obtained by filtration through a BeO filter. The spectrum of these neutrons is given in Fig. 3. Although these neutrons are not monochromatic they belong to a fairly narrow band of average wavelength 5·1Å. For the measurements at room temperature the gas under investigation was contained in a long aluminum tube of 2·5″ i.d. and 365 cm long. Pressures up to 2 atmospheres were used. On H_2 we performed some measurements at liquid air temperature and in this case a tube 24″ long was used. A copper coil was soldered

TABLE II. CROSS SECTIONS FOR $\lambda = 5·1$Å (GAS AT ROOM TEMPERATURE)

Molecule	σ (observed)	Sum of the total cross sections of constituent atoms	σ calculated from classical interference theory
CO_2	24·5	13·0	24·8 or 4·1 for opposite phase
N_2O	57·8	34	55 or 41 for opposite phase
O_2	16·2	8·2	13·2
N_2	47·4	30	44·4
CF_4	41·5	21	38 or 7·5 for opposite phase
H_2	170	42	

around this tube and liquid air was circulated in it. The entire assembly was protected with rock wool insulation. The conditions were such that the ortho-para ratio was practically the same as at room temperature.

Only in the cases of CF_4 and H_2 measurements were performed also for neutrons of shorter wavelengths obtained by Bragg reflection on a fluorite crystal.

In Table II the total cross sections of BeO-filtered neutrons observed for various gases at room temperature are collected. In the third column of the table is given the sum of the total cross sections of the constituent atoms. One notices considerable differences between this sum and the observed cross sections.

Except for the case of hydrogen which will be discussed later these differences are explained only to a small extent by the fact that the gas molecules are not at rest (so-called Doppler effect), and are due mostly to the interference of the waves scattered by the constituent atoms. In column 4 the values of σ calculated from the classical interference theory (see above) are given. For diatomic molecules containing two different atoms one obtains two different results, depending on the assumption that is made for the sign of the scattering lengths of the two atoms. In the cases of CO_2 and CF_4 a reasonably good agreement between theory and observation is obtained only if the scattering lengths

TABLE III. CROSS
SECTION OF CF_4
(ROOM TEMPERATURE)

$\lambda \times 10^8$	Velocity of neutrons (m/sec.)	$\sigma \times 10^{24}$
·881	4590	19·5
1·099	3600	19·0
1·316	3000	21·5
1·492	2650	20·8
5·1	775	41·5

of the component atoms are assumed to have the same sign. The same seems to be true for N_2O, though the sensitivity to a change in relative sign of the scattering length is not so pronounced here.

The very large cross section observed for H_2 is strongly perturbed by the Doppler effect because here the thermal agitation velocity of the molecules is larger than the velocity of the neutrons, so that the phenomenon can be better described by saying that the molecules hit the neutron rather than the opposite. This fact is brought out by the strong temperature dependence of the cross section which drops from 170×10^{-24} cm^2 to 81×10^{-24} cm^2

when the H_2 is cooled from room temperature to liquid air temperature. This point will be discussed later.

Table III gives the cross sections observed for CF_4 at room temperature for neutrons of various velocities. The table shows

TABLE IV.   CROSS SECTION OF H_2 AT ROOM TEMPERATURE

$\lambda \times 10^8$	Velocity (m/sec.)	$\sigma \times 10^{24}$	Factor for Maxwell distribution	σ/factor
·76	5212	50·9	1·048	48·6
·93	4256	53·2	1·071	49·7
1·11	3546	56·2	1·099	51·1
1·46	2714	61·2	1·172	52·2
1·68	2347	64·4	1·231	52·3
2·24	1765	68·0	1·407	48·3
5·1	775	169·7	2·514	67·5

TABLE V.   CROSS SECTION OF H_2 AT 83°K (ORTHO-PARA RATIO UNCHANGED FROM ROOM TEMPERATURE)

$\lambda \times 10^8$	Velocity (m/sec.)	$\sigma_{\text{total}} \times 10^{24}$ observed	$\sigma_{Sc} \times 10^{24}$ corrected for absorption and Doppler effect	Teller-Schwinger theory
0·836	4730	44	43	—
1·202	3290	51	49	—
1·638	2420	55	52	56
1·884	2100	56	51	57
5·1	775	81	52	80

a rise by about a factor 2 when the wavelength increases from 1·5 to 5·1Å. Up to 1·5Å the cross section is fairly close to 21×10^{-24} cm^2, namely to the sum of the cross sections of one atom of carbon and four of fluorine. The rise is due in large measure to interference.

Tables IV and V give the results obtained on H_2 at room temperature and at liquid air temperature respectively. At room

K*

temperature one notices a steady rise in cross section with decreasing velocity of the neutrons. As previously indicated, this is to a large extent due to the thermal agitation velocity of the gas molecules. A correction factor, calculated on the assumption that the cross section is constant, is given in the fourth column. The last column gives the cross section corrected by this factor.

The data of Table V taken at liquid air temperature can be most easily compared with the Teller-Schwinger theory because at this temperature practically only rotational states 0 and 1 are involved. Column 3 of this table gives the observed cross section and column 4 gives the cross section corrected for the thermal agitation velocity of the gas molecules. The fifth column gives the theoretical values calculated with the Teller-Schwinger theory, assuming $a_0 = -2.40 \times 10^{-12}$ and $a_1 = 0.54 \times 10^{-12}$ for the singlet and triplet state of the neutron proton system. The agreement is not too good especially for the lowest velocity, though the last value may be vitiated by the large Doppler correction.

By performing a similar set of measurements on deuterium one might be able to draw some conclusions of the spin dependence of the cross section, and we hope to make such measurements in the future.

5. *Total Reflection of Neutrons on Mirrors*

The total reflection of neutrons on mirrors is theoretically expected to occur at very small glancing angles for substances with a positive scattering length. The index of refraction is given by

$$n = 1 - \lambda^2 Na/2\pi \qquad (9)$$

where a is the scattering length taken with the proper sign, N is the density of atoms in the mirror, and λ is the wavelength. Consequently n is less than 1 (case of total reflection) if a is positive. The limiting glancing angle is then given with very good approximation by

$$\theta_0 = [2(1-n)]^{\frac{1}{2}} = \lambda(Na/\pi)^{\frac{1}{2}}. \qquad (10)$$

In most cases $(n-1)$ is of the order of magnitude of 10^{-6} and the limiting angle of the order of 10'. The total reflection of neutrons on mirrors has been previously observed[4] using non-monochromatized thermal neutrons. In this case of course one cannot observe a limiting angle because θ_0 [see (10)] is proportional to λ so that neutrons of the various wavelengths of the Maxwell distribution drop out of the reflected beam gradually as the glancing angle is increased. In spite of this, the fact that strong reflection

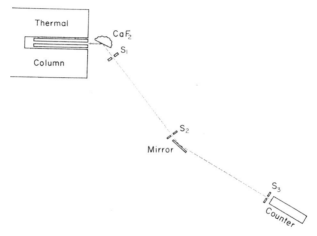

Fig. 4. Monochromatic total reflection on mirrors.

is observed on a substance can be construed as good evidence that the index of refraction for that substance is less than one, and that therefore a is positive. This is so because, although a substance which is not totally reflecting has a finite reflection coefficient, it is so small that no very prominent reflection could be observed in such cases. From the quoted experiments it could therefore be concluded that the following elements, Be, Cu, Zn, Ni, Fe, C, for which strong reflection has been observed, have positive scattering length.

[4]Fermi and Zinn, *Phys. Rev.* **70**, 103 (1946).

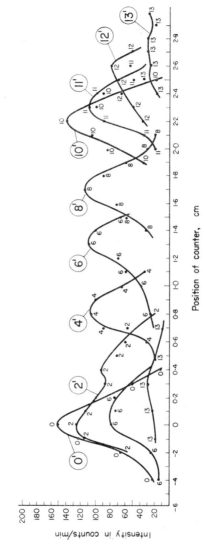

FIG. 5. Typical results of measurements of a reflection on a Be mirror.

In order to test conclusively the theory of total reflection, the reflection experiments were repeated using monochromatic neutrons. This made it possible to observe the sharp drop of the reflected intensity at the limiting angle and to measure this angle.

The experimental arrangement is shown schematically in Fig. 4. A collimated beam of thermal neutrons emerging from the thermal column is made monochromatic by Bragg reflection on a fluorite crystal. It is further collimated by passing through two very narrow cadmium slits, S_1 and S_2, 2 mm wide and about 3·5 meters apart. The narrow monochromatic beam emerging from the second slit falls on the mirror which can be rotated by small angles. The reflected beam is detected by a BF_3 proportional counter about 3·5 meters away and with a cadmium entrance slit 2 mm wide. By moving the counter across the beam, both direct and reflected beam are detected.

Typical results of these measurements are plotted in Fig. 5 for the case of a reflection on a Be mirror. The abscissa is the position of the detector expressed in centimeters measured transversely to the direction of the beam. The ordinate is the neutron intensity in counts per minute. The curves whose experimental points are marked 0, 2, 4, etc., correspond to the intensity plots found when the mirror is set at an angle of 0, 2, 4, etc., minutes to the direction of the neutron beam. When the mirror is parallel to the beam, there is observed a single maximum corresponding to that of the main beam. When the mirror is rotated by 2′ the reflected beam is barely resolved from the direct beam. At 4′ the two maxima are well separated. For larger angles the direct beam has not been drawn in the plot except for the largest angle, 13′, at which the reflected beam has completely disappeared and also the direct beam has disappeared because the mirror has turned by such a large angle that it completely cuts off the main beam. These points therefore correspond to the background, which actually was about 20 counts per minute.

By inspecting the plot one sees that the intensity of the reflected beam drops sharply to background between the positions 10 and 13 minutes. This drop is not quite sudden because the angular

TABLE VI. LIMITING ANGLE FOR TOTAL
REFLECTION OF NEUTRONS OF 1·873Å

Mirror	Limiting angle (minutes)	
	Observed	Calculated
Be	12·0	11·1
C (graphite)	10·5	8·4
Fe	10·7	10·0
Ni	11·5	11·8
Zn	7·1	6·9
Cu	9·5	9·5

TABLE VII. SCATTERING LENGTHS

Element	$a \times 10^{12}$	$4\pi a_{at}^2 \times 10^{24}$	$\sigma_{scat} \times 10^{24}$	Remarks
Ba	0·79	7·7	8	C
Be	·89	8·0	6·1	M
Br	·56	3·9	< 7	C
C	·67	4·8	4·8	A
Ca	·79	7·5	9·5	C
Cl	1·13	15	15	A, no spin and isotope effect considered
Cu	·81	8·0	7·2	M
F	·60	4·1	4·1	A, no spin effect considered
Fe	·82	8·1	9·2	M
H	−·39	·48	21	From theory
I	·36	1·6	1·6	A, no spin effect considered
K	·35	1·5	1·5	A, no spin and isotope effect considered
Li	−·59	3·4	—	C
Mg	·56	3·7	4	C
Mn	−·44	2·4	2·4	A, no spin effect considered
N	·87	8·3	8·3	A, no spin effect considered
Na	·56	3·7	3·5	C
Ni	1·09	14	13	M
O	·61	4·1	4·1	A
Pb	·48	2·9	10	C
S	·28	·95	1·1	C
Zn	·58	4·1	3·6	M

resolution was about two minutes. A study of the intensity shows the limiting angle to be 12 minutes.

In Table VI the values of the limiting angles of several mirrors measured for neutrons of de Broglie wavelength $\lambda = 1\cdot873\text{Å}$ are given. The experimental values given in the second column of the table should be compared with the theoretical values calculated from formula (10) given in the third column. The agreement is excellent except in the case of carbon, for which a not too good graphite mirror was used.

The absolute determination of the sign of the scattering length for the totally reflecting elements, taken together with the results of Sections 2 and 4, allows one to determine the actual sign for a large group of elements.

6. *Conclusions*

In Table VII all the data available at present on scattering length have been summarized. The absolute sign given in the table has been obtained primarily from the data of Section 5 on total reflection, and indirectly, by combining these data with those of Section 2 on intensities of Bragg reflections, and Section 4, on molecular scattering. The absolute values have been obtained in various ways.

For the elements having only one isotope or one strongly dominant isotope, and no nuclear spin, the absolute value of the scattering length can be calculated from formulae (5) and (2) provided the total scattering cross section is known. For example, this method has been applied in assigning the value $0\cdot61 \times 10^{-12}$ cm for the scattering length of oxygen in a crystal. This differs by a factor 16/17 from the scattering length $0\cdot57 \times 10^{-12}$ cm for a free oxygen atom. In some cases this same method has been applied to elements having several important isotopes, or a spin, or both. In such cases the absolute value is naturally much less reliable. The method has been used when no better way was available. In the last column of Table VII the code letter A has been used to denote values obtained by this method.

The results of Section 2 allow the scattering lengths of several elements to be related directly or indirectly to that of oxygen whose scattering length 0.61×10^{-12} has been used as a standard for these cases.

For the five elements Na, K, Cl, Br and I, which are not conclusively related to the others (see Section 2), Cl has been taken as a standard with a scattering length of 1.13×10^{-12} cm. The sign has been taken positive partly on account of the evidence from Bragg reflection on $NaNO_3$, as discussed in Section 2. Furthermore, both theoretical and experimental evidence indicate that cases of negative scattering length are very improbable. The probability that five elements all having the same sign should be negative is therefore exceedingly small.

In other cases the value of the scattering length has been calculated from the value of the limiting angle using formula (10). Code letter M indicates these instances.

In the third column of Table VII is given the expression $4\pi a^2 (A/A+1)^2 \times 10^{24}$. If there are no important isotopes and no spin this should coincide with the scattering cross section. Otherwise it should in general be smaller, the more so the greater the differences are in the scattering lengths for the various isotopes and spin orientations.

The values in column 3 therefore should be compared with the experimental values of the scattering cross section given in column 4. Naturally the comparison is trivial for the cases designated by A, where formulae (2) and (5) have been used to calculate a. Comparison in the remaining cases shows that usually the data of the third and fourth columns are rather close. The few exceptions for which the data in column 3 are less than those of column 4 can be explained as due to experimental inaccuracy. The fact that they are close seems to indicate that, in general, the spin and isotope dependence of the scattering length is not extremely pronounced.

There are two notable exceptions. One is hydrogen, a case well known from the Teller-Schwinger theories. The other one is lead where perhaps some of the many isotopes may have a negative scattering length.

The table shows that the scattering length is positive in the great majority of cases as discussed in Section 1.

This work was performed at the Argonne National Laboratory, and we greatly benefited by the help of the staff of the laboratory in providing and operating the neutron source, and in supplying many of the experimental facilities needed. Several of the crystals used were loaned to us by the Field Museum in Chicago through the courtesy of Mr. Changnon, by Dr. Howland of the Minerology Department of Northwestern University, and by Dr. Fisher of the Geology Department of the University of Chicago. Also some specimens were obtained from the Union Carbide and Carbon Research Laboratories, Inc., and the Eagle Picher Lead Company. Mr. Warren Nyer helped in some of the experiments.

This document is based on work performed under Contract No. W-31-109-eng-38 for the Manhattan Project, and the information covered therein will appear in Division IV of the Manhattan Project Technical Series as part of the contribution of the Argonne National Laboratory.

Note by G. E. B. The use of the term "form factor" in this paper to describe the overall amplitude of scattering by a unit cell, as in equation (8) on page 275, is not in accordance with convention. The usual term is "structure amplitude factor", as in our discussion of equation (5.2) on page 45. The term "form factor" is usually reserved to describe the angular variation of scattering from a single *atom*, such as occurs for X-rays or for the *magnetic* scattering of neutrons which we discussed in equation (8.1) on page 76.

15 The Diffraction of Thermal Neutrons by Single Crystals*

G. E. BACON†

Experimental measurements of the intensity of neutron reflexion by large single-crystal slices are shown to be in agreement with previously published calculations. It is demonstrated that accurate comparison can be made of the structure factors of planes in a zone by the use of cylindrical or pillar crystals, sufficiently small in diameter to give negligible secondary extinction effects. The use of single crystals for structure analysis is discussed.

Introduction

Measurements of the intensity of reflexion of thermal neutrons by a large number of single crystals were published by Fermi & Marshall[1]. These workers found that the reflectivities were more nearly proportional to the crystal-structure factor than to its square, in contrast with the case for X-ray reflexion by mosaic crystals such as were employed. This result is accounted for by the predominant influence of secondary extinction on the reflexion of neutrons from thick mosaic crystals, as pointed out by Bacon & Thewlis[2] and considered in detail by Bacon & Lowde.[3] The latter writers showed theoretically that proportionality between reflectivity and the square of the structure factor can only be obtained for very small crystals and gave criteria for the permissible size of crystals. The purpose of the present paper is to present

---

*Proceedings of the Royal Society, A, **209,** 397–407 (1951).

†Atomic Energy Research Establishment, Harwell. Communicated by Sir John Cockcroft, F.R.S. Received 27 June 1951.

[1]E. Fermi and L. Marshall. *Phys. Rev.* **71,** 666 (1947).
[2]G. E. Bacon and J. Thewlis. *Proc. Roy. Soc.* A, **196,** 50 (1949).
[3]G. E. Bacon and R. D. Lowde. *Acta Cryst.* **1,** 303 (1948).

experimental evidence of these conclusions. It is shown that under certain circumstances single crystals, with their inherent advantages, can be used for neutron crystallography, thus extending the field of structure analysis beyond the limitations of the powder methods so far employed.

Theoretical relationships

The case treated by Bacon & Lowde is that of the symmetrical reflexion of a monochromatic neutron beam by a plane parallel plate of crystal. They show that the integrated reflexion $\mathscr{R}^{\theta}$ is given by

$$\frac{1}{\eta} \mathscr{R}^{\theta} = \psi \left(\frac{\mu t}{\sin \theta}, \frac{1}{\eta} \frac{Q}{\mu} \right),$$

where η is the standard deviation of the angular spread of the mosaic blocks

μ is the linear coefficient of true absorption

t is the thickness of the crystal

θ is the Bragg angle

Q is the crystallographic quantity given by

$$Q = \frac{N^2 F^2 \lambda^3}{\sin 2\theta},$$

where N is the number of unit cells per cm.[3]

F is the structure factor of unit cell for neutrons (see Note 2)

λ is the neutron wavelength.

The form of the function ψ, which will not be detailed here, is such as to give curves of the type shown in figure 1 in which $\frac{1}{\eta} \mathscr{R}^{\theta}$ is plotted as a function of $\frac{1}{\eta} \frac{Qt}{\sin \theta}$ for various values of $\mu t / \sin \theta$.

The principal features of these curves are the following:

(a) As $\dfrac{1}{\eta}\dfrac{Qt}{\sin\theta}$ increases there is an initial region near the origin where $\dfrac{1}{\eta}\mathcal{R}$ is equal to $\dfrac{1}{\eta}\dfrac{Qt}{\sin\theta}$, or, more simply the integrated

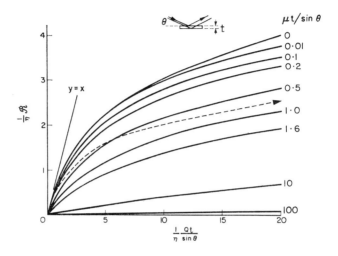

FIG. 1. Theoretical curves for the integrated reflexion from ideally imperfect crystal slices (reflexion case). Broken curve shows mean line through area of diagram explored by experiment.

reflexion is given by QV, where V is the volume of crystal being irradiated.

In the lowest curves this initial part close to $y = x$, which is a line inclined at 76° to the horizontal, is too small to be distinguished.

(b) With further increase of $\dfrac{1}{\eta}\dfrac{Qt}{\sin\theta}$, $\dfrac{1}{\eta}\mathcal{R}$ becomes *proportional* to this function in the case of the curves for large values of

$\mu t/\sin \theta$, which correspond to the case of *X-ray* diffraction and, in fact, $\mathscr{R}$ is equal to $Q/2\mu$, as is well known. For example, on the lowest curve, for which $\mu t/\sin \theta = 100$ the value of $\dfrac{1}{\eta} \mathscr{R}$ when $\dfrac{1}{\eta} \dfrac{Qt}{\sin \theta} = 20$ is 0·1, which is also found to be the value of $\dfrac{1}{\eta} \dfrac{Q}{2\mu}$.

(*c*) For the curves having smaller values of $\mu t/\sin \theta$, which are the ones in general applicable to *neutron* diffraction, the fall off from the line $y = x$ is followed by a region where $\mathscr{R}$ is no longer proportional to Q but is determined largely by the mosaic spread η of the crystal. Here secondary extinction is predominant and the value of $\mathscr{R}$ is of the order of 3η.

Experimental measurements

I. *Crystal slices.* The first test of the theoretical conclusions outlined above was made using plane-parallel plates of potassium bromide, 5 cm. square and ranging from 0·17 to 1·38 cm. in thickness and cut with their surfaces parallel to the (100) planes. A number of reasons determined the choice of this substance. In particular, the measured value of μ was 0·2 cm.$^{-1}$ at $\lambda = 1·25$Å, and this is a value typical of a large number of substances for neutron absorption; the unit cell size, $a = 6·59$ Å, is quite large resulting in several orders of reflexion falling within the restricted range of Bragg angle, θ, of the neutron spectrometer; the substance can be obtained in large plates of various thicknesses. The ranges of values of μ, t, θ, η used in the measurements are such as to give values of $\mu t/\sin \theta$ ranging from 0·06 to 1·45 and values of $\dfrac{1}{\eta} \dfrac{Qt}{\sin \theta}$ from 0·04 to 50. The combinations are such, however, as to concentrate interest in the portion of the area of figure 1 which is close to the chain dotted line shown there, assuming the line to be continued beyond the right-hand edge of the diagram to a value of about 3 for $\dfrac{1}{\eta} \mathscr{R}$ at a value of 50 for

$\dfrac{1}{\eta} \dfrac{Qt}{\sin\theta}$.

The measurements were made with the double-crystal neutron spectrometer described by Bacon, Smith & Whitehead,[4] the crystal under test being mounted on the second rotating table of this intrument, to receive a beam of monochromatic neutrons reflected from a calcium fluoride crystal on the first table. Both tables were surrounded by circular drums of borated paraffin wax to reduce the number of background neutrons scattered into the counter. The two drums were connected by a cadmium-lined tunnel, about $1\frac{1}{2}$ in. square in cross-section, the ends of which could be closed by cadmium plates having rectangular slits of various sizes for the purpose of restricting the dimensions of the neutron beam falling on the test crystal. Most of the measurements were made with beams either $\frac{1}{2}$ or 1 in. high and either $\frac{1}{16}$ or $\frac{1}{8}$ in. in width. Narrow beams of these widths were used in order to ensure that, even for the smallest values of θ employed ($\theta = 11°$), the 2 in. width of crystal would be adequate to receive the beam, thus approximating the theoretical case of the infinitely wide plane-parallel slab. In fact, separate measurements with the $\frac{1}{16}$ and $\frac{1}{8}$ in. beams were in agreement, within the experimental error, suggesting that the approximation was satisfactory. The crystals were mounted on a modified goniometer mount, with tilting and centring arrangements, and the counting of the reflected neutrons was done either point by point for successive positions of the counter arm and crystal table or continuously by automatic recording. In the latter case the counter arm rotated continuously at an angular speed of 1° of arc in 10 min. of time, the crystal following at half this speed. In either case the integrated reflexions were measured by the area, above background, under the diffraction peaks.

[4]G. E. Bacon, J. A. G. Smith and C. D. Whitehead. *J. Sci. Instrum.* **27**, 330 (1950).

Discussion of results. Figure 2 shows the experimental results for the 200, 400 and 600 reflexions from potassium bromide plates of three different thicknesses. The values of the integrated reflexions (all in the same arbitrary units) are plotted against $Q/\sin\theta$, where Q has been corrected for the effect of thermal vibrations assuming the usual value of 178° K for the Debye

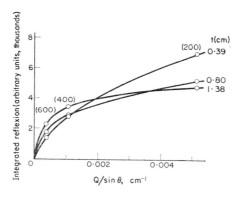

FIG. 2. Experimental measurements of (*h*00) reflexions for crystal slices of various thicknesses.

temperature. It will be shown, in turn, that both the shapes of the curves for the different thicknesses of crystal and their relative values for the various orders of reflexion are consistent with the theory.

In figure 3 the curves are replotted to simplify comparison of their shapes by normalizing the value of the (200) reflexion to unity in each case. In addition an experimental curve is added for an even thinner crystal plate ($t = 0.17$ cm.) which was measured under slightly different conditions which did not permit direct comparison with the individual values of $\mathscr{R}$ shown in figure 2. These four curves show the progressive change from the case of the thinnest crystal, for which $\mathscr{R}$ is practically proportional to $Q/\sin\theta$, or QV, over the range of measurement, to that of the thickest crystal for which secondary extinction is predominant and affects markedly even the weakest (600) reflexion, whose intensity is well below the value which would be given by the tangent to the curve at the origin. In fact, the onset of secondary extinction is accentuated by, as will be seen later, the smaller mosaic spread η of the thicker samples. Measurement of the value of η for samples of this sort is not easy; theoretically, it can be done by measuring the width of the rocking curve of the crystal when mounted in the "parallel position" for double-crystal

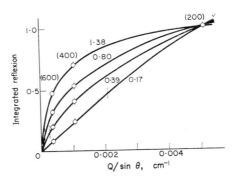

Fig. 3. Experimental intensities of (h00) reflexions normalized to (200). Numbers against curves give values of t in cm. At positions indicated thus

$$\left| \; \frac{1}{\eta} \frac{Qt}{\sin \theta} = \frac{1}{4} \right.$$

spectrometry using an identical crystal as monochromator (Compton & Allison, p. 718).[5] In practice, however, the rocking curves are broadened owing to the crystals not being truly single but consisting of two or three individuals over the portion covered by a wide initial beam, this being evidenced by the shapes of the rocking curves. From these curves it is concluded that the mosaic spread is of the order of a few minutes of arc and is certainly least for the thickest crystals; no more precise conclusion is

[5]A. H. Compton and S. K. Allison. *X-rays in theory and experiment.* New York: Van Nostrand, 1935.

felt to be justified. However, information on this point can be obtained by considering the experimental points shown in figure 3 in relation to theoretical curves drawn, for a given thickness of crystal, for different values of η. These curves are shown for two

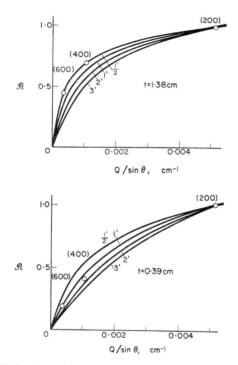

FIG. 4. Estimation of mosaic spread from experimental intensities. Numbers against curves show values of η.

of the crystals in figure 4. As would be expected, the thicker the crystal and the smaller the value of η, the more does the theoretical curve depart from linearity owing to the effect of secondary extinction. From figure 4 and the corresponding curves for the other two crystals it is estimated that the values of η for the four crystals are, respectively, 0·6′, 1′, 1·5′ and 5′. The appreciable increase for the thin plates does not seem unreasonable in view of the strains which must be imposed during the cleaving and working operations. In this connexion it may be worth emphasizing that the whole body of these low-absorbing crystals is contributing to the reflexion and it is not a case of surface reflexion,

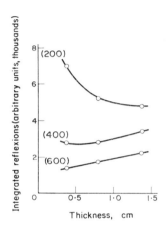

FIG. 5. Variation of experimental intensities with thickness.

as for X-rays. Consequently, when the mosaic spread of the surface layer of the thickest crystal was increased by roughening with sandpaper, no increased reflexion was observed experimentally; the extent to which secondary extinction limits the reflexion is determined by the mosaic spread throughout the volume of the crystal.

Confirmatory evidence supporting these deductions regarding mosaic spread is forthcoming from a consideration of the relative $\mathscr{R}$ values, of the various spectra, for the different thicknesses of crystal which figure 2 shows. In particular, it is clear from the figure that although there is an increase of $\mathscr{R}$ with thickness for

(600), yet there is a fall for the (200) reflexion, i.e. fewer neutrons are reflected from the thick crystal than from the thin one. In order to simplify the interpretation of this, at first sight, surprising result, the experimental points are replotted in figure 5, where $\mathscr{R}$ for each reflexion is shown as a function of thickness. The explanation of these results then follows from figure 6 where, for

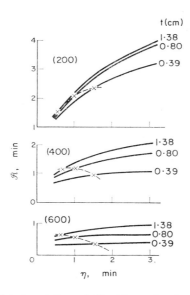

FIG. 6. Calculated variation of $\mathscr{R}$ with η, showing experimental positions.

each reflexion, are shown theoretical curves (derived from figure 1) giving the variation of $\mathscr{R}$ with η for the three thicknesses of crystal. In the case of the weakest reflexion, (600), $\mathscr{R}$ is largely determined by thickness and is little influenced by η. On the other hand, for the much more intense (200) reflexion η is much more important than t. Superimposed on the curves for each reflexion is a dotted curve which links the values of mosaic spread which were estimated above for each thickness of crystal. The corresponding values of $\mathscr{R}$ show the same kind of variation with thickness as the experimental curves in figure 5, supporting the conclusion of variation of mosaic spread with thickness. Finally, figure 7 summarizes the general agreement of proportionality

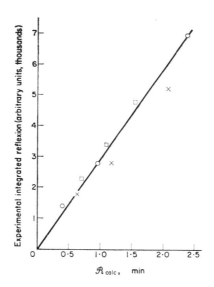

FIG. 7. Crystal slices: comparison of experimental and calculated intensities. $t = \odot$, 0·39 cm.; $\times$, 0·80 cm.; $\square$, 1·38 cm.

between the nine experimental points, from these three crystals, and the theoretically calculated integrated reflexions using the estimated values of η.[6]

It was shown by Bacon & Lowde that for a non-absorbing crystal $\dfrac{1}{\eta}\mathscr{R}$ is within 5% of the value of QV, thus giving propor-

---

[6]Later measurements have placed the experimental intensities on an absolute scale. They show that the mean line in figure 7 gives agreement between theoretical and experimental values to within 10%.

tionality between $\mathscr{R}$ and Q, over the range of variables for which $\dfrac{1}{\eta} \dfrac{Qt}{\sin\theta} < \dfrac{1}{4}$. In the present case the values of $Q/\eta\mu$ are rather small for the crystal to be regarded as "non-absorbing", but the criterion still gives an approximate idea of the range of proportionality. This can be seen from figure 3, where the dashed vertical line on each curve marks the position at which $\dfrac{1}{\eta} \dfrac{Qt}{\sin\theta} = \dfrac{1}{4}$.

II. *Cylindrical or pillar crystals.* The measurements so far described have shown that, under the conditions considered, it is possible to use crystal plates sufficiently thin ($t = 0\cdot17$ cm.) to give approximate proportionality between $\mathscr{R}$ and Q for successive orders of reflexion from a crystal slice. With a view to exploring further the possibilities of single-crystal analysis the measurements were then extended to a case of more practical interest, that of a square pillar of crystal which could be bathed in the neutron beam (ideally a cylinder would have been more convenient).

Initially a pillar about 4 mm. square and $1\frac{1}{2}$ cm. high was used, mounted with its length vertical and bathed in a neutron beam about 7 mm. in width. The length of the crystal was parallel to a [100] axis, and it was possible to measure a full zone ($hk0$) of reflexions instead of merely the successive orders of ($h00$). The results are shown in figure 8, in which $\mathscr{R}$ is plotted directly against Q, corrected for thermal vibrations. The curvature of the mean line through the points is found to be less than for a plane-parallel slab of thickness 4 mm., equal to the pillar dimension; this is as would be expected, for the effect of secondary extinction is determined by the distance of travel of the neutrons within the crystal, and when θ is small, as for (200), this distance for the pillar will be smaller than $t/\sin\theta$ which is the value in the case of the parallel slab.

In figure 9 the results are shown for a fragment of crystal from the thinnest slice used in the earlier work. This prismatic fragment was slightly under 2 mm. square, and a length of 1 cm.

was bathed in the neutron beam. $\mathscr{R}$ is now proportional to Q over practically the whole range of measurement, as a result of the reduction of the neutron path length through the crystal and the simultaneous increase of η.

It becomes of interest to consider methods of increasing the

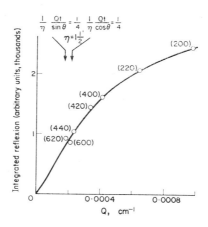

FIG. 8. Intensities of $(hk0)$ reflexions from pillar of crystal 4 mm. square.

mosaic spread throughout the volume of these crystals, since the maximum permissible thickness is directly proportional to η. It is well known that, in some cases at least, an increase of η can be obtained by subjecting a crystal to a number of immersions in liquid air. Figure 10 illustrates the possibilities of this, the full line showing the experimental curve obtained after such treatment of the 4 mm. square crystal used above for figure 8. Comparison with the dotted curve, obtained before the liquid-air treatment, shows that a marked improvement in linearity has been achieved.

General conclusions: neutron crystallography

From the results which have been described it is concluded that, for the material used, proportionality between $\mathscr{R}$ and Q is obtained for the successive reflexions from a parallel crystal slice of thickness $t = 0.17$ cm. With thicker crystals it has been shown that secondary extinction causes the integrated reflexion to fall away from this proportionality to an extent in agreement with the calculations of Bacon & Lowde (1948). Similarly, by using a vertical square pillar of crystal with horizontal edges of this same dimension proportionality is obtained for all the reflex-

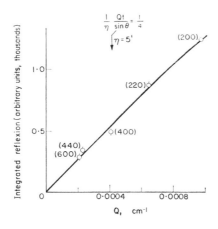

FIG. 9. Intensities of $(hk0)$ reflexions from pillar of crystal 2 mm. square.

ions in the $(hk0)$ zone; moreover, by increasing η by liquid-air treatment, similar results could be achieved for the 4 mm. square crystal.

These dimensions of a fraction of a centimetre, in the present case, suggest that the estimate of a maximum crystal thickness of a fraction of a millimetre given by Bacon & Lowde was unduly pessimistic; however, the reason for their much lower value lies in their use of $Q = 0.02$ cm.$^{-1}$ as an average value of Q for an intense reflexion. This was the value, at $\lambda = 1.5$ Å, for the (200) reflexion of magnesium oxide, a substance for which earlier

theoretical calculations had been made. However, a large number
of substances of practical interest will have maximum Q values
much smaller than this, for the expression for Q contains the
reciprocal of the unit cell dimension, 'a', raised to the sixth
power. Magnesium oxide, for which $a = 4.2$ Å compared with
values of about 6 Å for alkali halides, has a high value of scattering

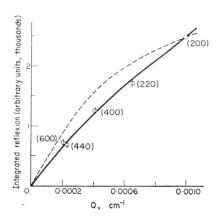

FIG. 10. Effect of increasing mosaic spread by immersion in liquid
air. – – –, form of original curve; ———, after liquid air
immersion.

power per unit volume. Further, Q can always be reduced by
reducing the neutron wavelength from 1·5 to 1·0 Å. In the present
case it was convenient to work with a wavelength of 1·25 Å, and
Q for (200), the most intense reflexion for potassium bromide, is
only 0·001 cm.$^{-1}$. Consequently the thickness of crystal for which
secondary extinction is negligible is proportionately greater.

Many substances concerning which useful information could
be obtained by neutron diffraction will have maximum Q values
no greater than this. It is felt therefore, in light of the above
results, that more consideration should be given to the use of

single-crystal methods. Better resolution can be obtained than by using powder methods, overlapping of spectra is avoided and the difficulties caused by incoherent background scattering are much reduced. The latter follows, since when a single crystal is used the whole crystal contributes to the peak, as well as to the background, whereas with a powder only a fraction of the material contributes to the coherent peaks but the whole contributes to the background. Important information concerning atomic positions of light elements should be forthcoming by using the expedient of a vertical pillar of crystal for measurement of a zone of reflexions. In the experiments described the pillar $0.2 \times 0.2 \times 1.0$ cm. gave a peak counting rate of about 70 counts per min. for (600), the weakest reflexion used. Intensities smaller by a factor of three could still be measured fairly accurately so that the use of smaller crystals is quite possible; with increased neutron fluxes it seems that the use of crystals more nearly spherical in shape, small enough to give negligible secondary extinction for reflexions in all directions, should be practicable. Moreover, by the use of monochromators of larger mosaic spread than the calcium fluoride crystal used in the above experiments, an increased neutron intensity in the monochromatic beam can be obtained. Preliminary measurements with a lead crystal, kindly supplied by Dr C. G. Shull of Oak Ridge and cut obliquely to the reflecting planes to give a reflected beam of reduced width, show an intensity increase of two or three times. These investigations are being continued to determine the miniumum size of crystals which will give adequate counting rates; at the same time attempts are being made to reduce the background count by using collimators of smaller aperture, appropriate to the small crystals being irradiated, rather than the large aperture collimators originally designed for powder diffraction.

This paper is published by permission of the Director of the Atomic Energy Research Establishment. The author is greatly indebted to Mr. R. F. Dyer for assistance with the experimental work and for carrying out practically the whole of the neutron counting.

L

16   A Single Crystal Neutron Diffraction Determination of the Hydrogen position in Potassium Bifluoride*

S. W. PETERSON and HENRI A. LEVY†

Neutron diffraction measurements on KHF_2 single crystals show that the hydrogen atom occupies the central position, within 0·1Å, in the linear F—H—F ion. The data also indicate asymmetry in thermal motion, which suggests that the bifluoride ion undergoes rotatory oscillation with appreciable amplitude. The study demonstrates the usefulness of single crystal neutron diffraction data for crystal structure determination.

Introduction

The question of the nature of the hydrogen bond in bifluoride ion has attracted much experimental and theoretical treatment. X-ray crystallography has shown that the two fluorine atoms in potassium bifluoride are separated by the short interatomic distance of 2·26Å. The location of the proton which is presumed responsible for the binding, has been the aim of several investigations. If the forces in the bifluoride ion are chiefly electrostatic, the proton would be expected to be centrally situated between the fluorine atoms. If exchange forces are important, a potential function with a double minimum might be expected and the proton probably would be situated at random in the two positions

Journal of Chemical Physics, **20,** 704–7 (1952). This work was performed for the AEC at the Oak Ridge National Laboratory.

†Chemistry Division, Oak Ridge National Laboratory, Oak Ridge, Tennessee. Received December 10, 1951.

close to one or the other fluorine atom. Ketelaar[1] and Glockler and Evans,[2] utilizing infrared data, came to the conclusion that the double minimum model was the correct one; their respective calculations led to distances of 0·70Å and 0·52Å between the potential minima. Some doubt was cast upon these conclusions by the dieletric constant measurements of Polder[3] and the calculations of Davies.[4] More recently, in a complete thermodynamic study, Westrum and Pitzer[5] showed that there is no residual entropy at absolute zero in the KHF_2 crystal and no specific heat anomaly above 15°K, strongly supporting the single potential minimum. In addition, they suggested a new assignment of the infrared frequencies consistent with their conclusion. This reinterpretation has been corroborated by Ketelaar and Vedder,[6] and by Newman and Badger,[7] who have recently reported polarized infrared studies of KHF_2, a method which allows increased certainty in assignment of frequencies. The present neutron diffraction investigation was begun before the appearance of the three latter papers. Even though the works of Westrum and Pitzer seem decisive, it is desirable to have confirmation of their conclusions by the essentially more direct method of a diffraction study. The result, which incidentally provides more detailed knowledge of the thermal motion of the bifluoride ion than was previously available, is presented below.

Procedure

X-ray studies [8,9] indicate that KHF_2 is tetragonal with $a_0 = 5·67$, $c_0 = 6·81$, of space group $D_{4h}^{18} - I\,4/mcm$, with

[1]J. A. A. Ketelaar, *Rec. trav. chim.* **60**, 523 (1941); *J. Chem. Phys.* **9**, 775 (1941).
[2]G. Glockler and G. E. Evans, *J. Chem. Phys.* **10**, 607 (1942).
[3]D. Polder, *Nature*, **160**, 870 (1947).
[4]M. Davies, *J. Chem. Phys.* **15**, 739 (1947).
[5]E. F. Westrum, Jr. and K. S. Pitzer, *J. Am. Chem. Soc.* **71**, 1940 (1949).
[6]J. A. A. Ketelaar and W. Vedder, *J. Chem. Phys.* **19**, 654 (1951).
[7]R. Newman and R. M. Badger, *J. Chem. Phys.* **19**, 1207 (1951).
[8]R. M. Bozorth, *J. Am. Chem. Soc.* **45**, 2128 (1923).
[9]L. Helmholz and M. T. Rogers, *J. Am. Chem. Soc.* **61**, 2590 (1939).

four molecules to the unit, potassium occupying the fourfold positions[10] (a) and fluorine the eightfold positions (h); the one parameter for fluorine was evaluated as 0·1408 by Helmholz and Rogers,[9] giving an F−F distance of 2·26Å. If the hydrogen is centrally situated in the bifluoride ion, it occupies the fourfold positions (d): $0\frac{1}{2}0$, $\frac{1}{2}00$, B. C. If asymmetrically situated, it can fit into the X-ray unit only if the orientation of the ion is random; with a colinear asymmetric ion, for example, $\frac{1}{2}H$ would occupy each of the eightfold positions (h): x, $\frac{1}{2}+x$, 0, etc. For a zero value of the parameter x, these positions reduce to the central positions (d). If the ion is asymmetric and the structure ordered, either an enlarged tetragonal unit or a departure from tetragonal symmetry is required.

The initial attack by neutron diffraction employed the usual powder method, with deuterated material to avoid heavy diffuse scattering due to hydrogen. The pattern was consistent in all respects with the X-ray unit, showing that the hydrogen positions do not call for an enlarged unit. An evaluation of the parameter x from the powder intensities was next undertaken. Although a zero value gave the closest approach, fully satisfactory agreement between calculated and observed intensities was not achieved. The difficulty was eventually traced to preferential orientation in the packed powder samples, with c axes tending to lie parallel to the vertical cylindrical axis of the sample container; this was clearly demonstrated when, in a test measurement, a pattern was obtained with the sample tilted at 90° to the normal orientation, and the intensity anomalies were reversed.

An attack by means of single crystal measurements was next undertaken. Recent work[11] has demonstrated that single crystal methods are applicable in neutron crystallography. The success achieved clearly confirms this conclusion and demonstrates that single crystal methods have many advantages over powder

[10] *International Tables for the Determination of Crystal Structures* (J. W. Edwards, Ann Arbor, Michigan, 1944), revised edition, Vol. 1, p. 226.

[11] S. W. Peterson and H. A. Levy, *J. Chem. Phys.* **19**, 1416 (1951); G. E. Bacon, *Proc. Roy. Soc.* (London) **209A**, 397 (1951).

methods in the neutron field as well as in the X-ray field. For this reason, the work will be reported in some detail.

For single crystal measurements a deuterated specimen is not required, because of the greatly enhanced ratio of intensity in a Bragg reflection to that in diffuse scattering. This arises from the fact that the entire single crystal specimen contributes simultaneously to a Bragg reflection. Hence, the ordinary hydrogen compound was used in the single crystal portion of this study.

Potassium bifluoride was prepared by direct reaction in platinum of aqueous KF and HF in equivalent quantity. Analysis by alkali titration of several preparations, dried to constant weight, showed less than 0·3 percent impurity. Several samples were deuterated and analyzed for deuterium in a manner to be described elsewhere[12] for use on collecting powder data. Single crystals of KHF_2 were grown by slow evaporation of aqueous solutions of the pure salt, yielding rectangular platelets up to nearly 2 mm in thickness (perpendicular to (c) and several millimeters on an edge bisecting (a) and (b)). A specimen approximately $1·4 \times 2·6 \times 3·6$ mm was selected for study.

The manner of collection of diffraction data from powder samples will be described elsewhere.[12] For single crystal data, the spectrometer was used with minimum collimation, the wavelength spread being of the order of $\pm 1·5$ percent at half-maximum intensity, and the sensitive angle subtended by the detector about 2°. The crystal specimen was mounted on a fine glass fiber set on a goniometer head and its orientation adjusted until alternately the [110], [010], [120], [230], and [001] zone axes were parallel to the spectrometer axis. Measurements were made with the crystal and detector rotating synchronously through the Bragg angle in the usual way, with a crystal angular speed of usually 8° per hour. The diffracted intensity was recorded automatically, as with powder samples. Simultaneously, the cumulative count across the reflection was recorded with a scaler. The integrated intensity was taken as the accumulated count

---

[12]H. A. Levy and S. W. Peterson (to be submitted for publication).

during the scanning, reduced by the background count as measured with the crystal out of reflecting orientation. Typical data for reflections at different scattering angles are shown in Fig. 1. Integrated intensities were easily reproducible to a few percent. The wavelength was 1·16Å.

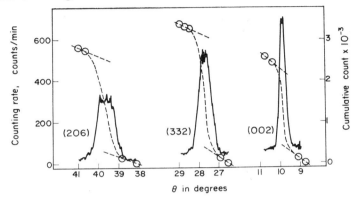

FIG. 1. Typical neutron diffraction reflections from a single crystal of KHF₂. The solid curves (left-hand ordinate scale) are traced from strip-chart records of the counting rate. The circled points (right-hand ordinate scale) show the observed cumulative count.

Treatment of data

Observed integrated intensities were corrected for absorption with a correction factor calculated as though the specimen were cylindrical with a cross section equal to its average cross section for the orientation under consideration. The linear absorption coefficient was measured on a sample of KHF₂ powder. In the case of reflections taken with the [001] zone vertical, which brought the maximum dimensions of the specimen into the equatorial plane, an extinction correction was necessary also. Only the strong (310) and (400) planes were appreciably affected. The latter was measured also in the (010] zone and served to evaluate the extinction coefficient in the Darwin expression.[13]

¹³See, for example, R. W. James, *The Optical Principles of the Diffraction of X-Rays* (G. Bell and Sons, Ltd., London, 1948), p. 292.

Because of uncertainties involved in this correction and greater importance of absorption, data taken with the [001] zone vertical were not used for parameter determination.

The tetragonal symmetry of KHF_2 was confirmed by establishing the equality of neutron intensities of several pairs of reflections of type $(hk0)$, $(kh0)$ with the [001] zone vertical. The possibility of ordered asymmetric structures was thus eliminated with reasonable certainty.

Structure factors $|F|$ were obtained from the corrected integrated intensities E by application of the following expression adapted from the standard X-ray expression[14]

$$E = \frac{I_0 \lambda^3 N^2 V}{\omega} \frac{|F|^2}{\sin 2\theta},$$

I_0 being the incident intensity, λ the wavelength, ω the angular velocity of rotation of the crystal, N the number of unit cells per unit volume, V the volume of the specimen, and θ the Bragg angle. The value of I_0 was established from similar measurements on a crystal of NaCl which had previously been carefully studied and shown not to be subject to appreciable extinction.[11]

Structure factors for models incorporating symmetric ions and linear asymmetric ions in disorder were calculated in the usual way. The nuclear scattering amplitudes were those reported by Shull and Wollan,[15] namely, $f_K = 0.35$, $f_F = 0.55$, and $f_H = -0.40$. The fluorine parameter value was that established by X-ray studies, 0.1408. The calculation was carried out for several values of the parameter x governing the degree of asymmetry. It was clear that use of a single Debye-Waller temperature factor would not bring satisfactory agreement with experimental values for any value of x. Separate temperature factors for K, H, and F were then introduced, and systematic variation led to much improved agreement. Careful examination of the comparison

[14]Reference 10, Vol. 2, p. 560.

[15]C. G. Shull and E. O. Wollan, *Phys. Rev.* **81**, 527 (1951). Hughes, Burgy, and Ringo, *Phys. Rev.* **77**, 291 (1950), have reported the somewhat smaller value $f_H = -0.38$. This value does not affect our conclusions in the current study.

at this stage disclosed that planes with large l index were yielding calculated structure factors consistently too large in absolute value, suggesting the need for an asymmetric temperature factor. No improvement resulted from the application of an asymmetric factor to K or H; however, the application to F brought about final satisfactory agreement. The form of the factor was

$$\exp[-B_F(\sin\theta/\lambda)^2 - B_F'(l/2c)^2],$$

TABLE I.   OBSERVED AND CALCULATED NEUTRON STRUCTURE.
FACTORS FOR KHF$_2$

hkl	$|F|_{obs}$	F_{calc} $x=0$	$x=\cdot02$	hkl	$|F|_{obs}$	F_{calc} $x=0$	$x=\cdot02$
110	1·16±0·05	1·16	1·13	332	2·70±0·11	−2·64	−2·77
002	1·37±0·06	1·30	1·30	215	2·10±0·10	2·12	2·15
200	0·93±0·06	−0·96	−0·92	422	1·45±0·08	−1·49	−1·35
112	1·59±0·07	−1·50	−1·52	006	0·53±0·05	0·54	0·54
211	3·08±0·10	3·06	3·10	116	1·21±0·08	−1·19	−1·20
202	3·40±0·10	−3·43	−3·39	334	0·38±0·11	−0·22	−0·32
220	< 0·19	0·10	0·18	206	2·39±0·10	−2·34	−2·31
222	2·39±0·09	−2·35	−2·27	424	0·59±0·12	0·71	0·83
004	3·49±0·11	3·44	3·44	440	2·56±0·10	2·62	2·81
213	2·64±0·09	2·70	2·74	325	1·10±0·07	−1·14	−1·20
114	1·10±0·07	1·05	1·03	226	1·67±0·09	−1·67	−1·62
321	1·60±0·08	−1·65	−1·76	442	0·35±0·20	0·52	0·69
204	0·62±0·08	−0·59	−0·55	600	1·80±0·06	1·75	1·94
400	3·28±0·11	−3·15	−3·01	062	< 0·27	0·22	−0·04
330	0·53±0·08	−0·44	−0·58	046	3·32±0·12	−3·41	−3·33
402	5·04±0·15	−5·23	−5·10	008	1·99±0·09	1·97	1·97
224	0·15±0·08	0·23	0·30	064	1·48±0·09	1·49	1·64
323	1·48±0·08	−1·46	−1·55	550	1·21±0·09	1·19	1·01
420	0·79±0·07	0·73	0·88				

in which B_F represents the effect of ordinary spherically symmetrical temperature motion, and B_F' the effect of an asymmetric component in the c direction. This temperature factor is similar to that applied to the CN group in NH$_4$CN by Lely and Bijvoet.[16]

[16]J. A. Lely and J. M. Bijvoet, *Rec. trav. chem.* **63**, 39 (1944).

Final values of the temperature factor coefficients were $B_K = 1\cdot25$, $B_H = 3\cdot0$, $B_F = 1\cdot85$, and $B_F' = 1\cdot0$.

Table I lists experimental structure factors for 37 reflections examined in the equators of the [110], [230], [120], and [010] zones of the single crystal specimen, together with estimated uncertainties in these values. Also listed are calculated structure factors for the symmetrically placed hydrogen ($x = 0$) and for the asymmetric random case corresponding to $x = 0\cdot02$. The

TABLE II.   NEUTRON DIFFRACTION DATA FOR EQUATORIAL
REFLECTIONS IN THE [001] ZONE OF KHF_2

| | Intensity, counts/128 | | | |
$hk0$	Obs.	Corrected*	F_{obs}	F_{calc} ($x=0$)
310	70·2	110	$4\cdot62\pm0\cdot20$	4·64
400	34·0	45·5	$3\cdot28\pm0\cdot15$	−3·15
510	17·6	22·4	$2\cdot52\pm0\cdot12$	2·50
530	4·0	4·4	$1\cdot16\pm0\cdot10$	1·03
620	0·3	0·3	$0\cdot3 \pm0\cdot20$	0·04
710	<0·5	<0·5	<0·4	0·01
640	3·7	4·1	$1\cdot14\pm0\cdot10$	−0·72
730	23·8	26·5	$2\cdot87\pm0\cdot30$	2·90

*Corrected for absorption and, in the case of (310) and (400), for extinction. The latter, together with the appropriate value from Table I, was used to establish the magnitude of the extinction correction.

same values of the temperature factors, listed above, were used in both calculations. Similar data taken in the equator of the [001] zone are collected in Table II. The symmetric model gives excellent agreement, while the asymmetric one gives agreement definitely less than satisfactory. The effect of change of temperature factors, particularly B_H, on the asymmetric model was carefully studied: the majority of the disagreements in column 4 of Table I can be made to approach agreement if a smaller value of B_H is used, for example, (222) and (400); however, the disagreement of (321), (213), and (325) is left essentially unchanged and the good agreement of a large number of other reflections,

L*

for example (226), (006), and (008), is rapidly destroyed by this change. We conclude that the asymmetric random model with $x = 0.02$ cannot give agreement with the data for any values of the temperature factor coefficients. Careful study of intermediate parameter values, with similar consideration of temperature

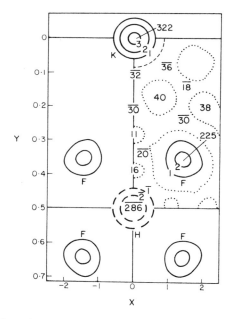

Fig. 2. A projection on the $a-b$ plane of the neutron scattering density in KHF_2. The figures indicate the values of the function at the contours and the extremes in units of 10^{-12} cm/$Å^2$. The dashed lines represent zero contours.

factor changes, places the upper limit of x at about 0.01. This modest precision, in view of the excellence of the agreement, is somewhat disappointing. It results primarily from the rather large temperature and zero point motion of the hydrogen atom, which reduces its contribution to the high angle reflections.

Broadening of these high angle reflections due to wavelength spread in the primary neutron beam, with attendant reduction of peak to background ratio and loss of precision in intensity measurement, also contributes.

The $(hk0)$ data were used to construct a projection of the neutron scattering density on the $a-b$ plane, shown in Fig. 2. The hydrogen atom, because of its negative amplitude, appears as a well rather than a peak on the map. Effects of termination of the Fourier series are apparent, particularly around the potassium atom, which has the smallest temperature factor; these could probably be reduced by application of a convergence factor to the data. The map is presented as an illustration of the possibility of applying series methods in neutron diffraction crystal analysis.

The powder data from KDF_2, after allowance for effects of preferred orientation, are fully consistent with the final model.

Discussion

This investigation has shown that the hydrogen position is within 0.1Å of the center of the $F-H-F$ link. This may be taken as confirmation of Westrum and Pitzer's result that the bifluoride ion is symmetric.

The root-mean-square displacements of the atoms corresponding to the temperature factors found are for K, 0.22Å; for H, 0.34Å; and for F, 0.29Å. It seems appropriate to interpret the asymmetry of the fluorine displacement in terms of components, perpendicular to and along the axis of the bifluoride ion. If it is assumed that displacements perpendicular to the ion axis have cylindrical symmetry, the data yield 0.10Å and 0.27Å, respectively, for the mean displacements parallel and in the perpendicular plane. The perpendicular displacement suggests a rotatory oscillation of the bifluoride ion with a mean half-amplitude of some $14°$. The relatively high displacement for hydrogen is not unusual; it probably represents mostly zero-point motion. The absence of appreciable asymmetry for hydrogen is consistent with recent assignments of spectroscopic frequencies which make the bending and asymmetric stretching frequencies not greatly different.[6,7]

The excellence of the agreement confirms previous reports[11] that single crystal neutron data are useful in favorable cases for structure determination. The absence of extinction in nearly all the measurements is consistent with Bacon and Lowde's[17] criterion; a rough application of their theory to the strong (211) reflection, assuming the low value of 5 minutes for the mosaic spread parameter, yields a limiting thickness of 2·1 mm, not greatly different from the smaller dimensions of our specimen. In the case of reflections measured in the equator of [001], where extinction appeared appreciable, the cross section of the specimen was 2·6 × 3·6 mm, which might well exceed the limiting value.

[17] G. E. Bacon and R. D. Lowde, *Acta Cryst.* **1**, 303 (1948).

17 On the Magnetic Scattering of Neutrons*

F. BLOCH

The direct experimental evidence of the neutron, obtained so far, indicates its mass and the range of forces within which it interacts with other heavy particles. The angular moments of nuclei make it practically sure that it has an angular momentum $\frac{1}{2}h/2\pi$. Furthermore there are good theoretical reasons to believe that it should have a magnetic moment of the same order of magnitude as the measured moment of the proton but having the opposite direction with respect to the angular momentum; these conclusions are partly based on Fermi's theory of the β-decay, partly on the known magnetic moment of the deuteron. Since the Stern-Gerlach method may meet considerable difficulties when applied to neutron beams, we want to propose a different way of obtaining information about the magnetic moment of the neutron which seems considerably simpler and promising in several other respects.

Consider an atom (or molecule) which in its ground state has a total magnetic moment μ caused by the spin or the orbital motion of the atomic electrons. The magnetic field around and within the atom can in any case be described by an average dipole density distribution $\mu g(\mathbf{r})$ with $\int g(\mathbf{r})dr = 1$. It will scatter neutrons on account of two reasons:

(1) Because of the interaction of the neutron with the atomic nucleus (or nuclei);

*Physical Reveiw, **50**, 259–60 (1936).

(2) Because of the inhomogeneous magnetic field in its surrounding acting on the magnetic moment of the neutron.

Although the forces on the neutron due to the second cause have to be assumed to be extremely much weaker than those due to the first cause, they act on distances so much larger that the scattering effect of both on slow neutrons becomes of the same order of magnitude. Treating the interaction due to both causes as small disturbances of the plane waves, which represent the incoming and scattered neutron, one readily obtains a formula for the magnetic influence on the scattering process.

Let θ be the angle between the orientation of μ and the direction of incidence of a neutron with velocity v, $\gamma_n = \mu_n/[(e/Mc)\cdot(h/4\pi)]$ the magnetic moment of the neutron μ_n, measured in units of the Bohr magneton, divided by the ratio of masses M/m of the neutron and electron and $\mathbf{q} = \mathbf{k}_0 - \mathbf{k}_1$ the difference between the vectors of propagation of incident and scattered wave, both having equal magnitude $k_0 = k_1 = 2\pi Mv/h$. The cross-section ϕ_ω per unit solid angle for scattering under an angle ϑ against the direction of incidence and an azimuth φ against the common plane of μ and $\mathbf{k}_0$ is then given by

$$\phi_\omega = \sigma_\omega \left| 1 \pm \frac{\gamma_n \gamma_e}{2(\sigma_\omega)^{\frac{1}{2}}} \frac{e^2}{mc^2} \left(\sin\theta\cos\frac{\vartheta}{2}\cos\varphi - \cos\theta\sin\frac{\vartheta}{2}\right)^2 F(\mathbf{q}) \right|^2, \tag{1}$$

where γ_e is the absolute magnitude of the atomic moment μ, measured in units of the Bohr magneton, and

$$F(\mathbf{q}) = \int \exp{(i(\mathbf{q}\cdot\mathbf{r})g(\mathbf{r}))}dr \tag{2}$$

is an atomic form factor, determined by the distribution of magnetism in the atom, which approaches unity for $1/q$ being large compared with atomic dimensions. The plus or minus sign in formula (1) is valid for neutrons with a magnetic moment oriented parallel or antiparallel to μ, respectively.

Formula (1) for the scattering cross section per atom remains practically valid also for the case of a ferromagnetic poly-crystalline substance, the only difference being that for the determination of **q** only such neutron velocities v are to be used for which the condition of interference at microcrystals with properly chosen orientation can be satisfied. Furthermore one has to consider that γ_e becomes temperature dependent:

$$\gamma_e(T) = \gamma_e(0)\,\frac{I(T)}{I(0)} \tag{3}$$

[$I(T)$ = Intensity of magnetization at saturation and at absolute temperature T] because of the decreasing average magnetization per atom as the temperature T approaches the Curie point; at saturation the angle θ in (1) is the angle between the magnetizing external field and the direction of incidence of the neutrons. While for fast neutrons the second term in (1) is negligible, it is quite considerable for neutrons with thermal energy, for which the wavelength is comparable with atomic dimensions, since $F(\mathbf{q})$ has then the order of magnitude one. The importance of the magnetic effect is measured by the number $k = (\gamma_n\gamma_e/2(\sigma_\omega)^{\frac{1}{2}} \times$ $\times (e^2/mc^2)$ which, for example for magnetized iron with $\gamma_e \cong 2$, $(\sigma_\omega)^{\frac{1}{2}}$ = radius of the iron nucleus = 5.10^{-13} cm and assuming $\gamma_n = 1$ becomes $k \cong 0.7$.

Since for $T >$ Curie temperature γ_e vanishes, one can obtain for temperatures of the ferromagnetic scatterer above the Curie point independent information about σ_ω alone. For temperature below the Curie point the effect then depends solely on the product $\gamma_n F(\mathbf{q})$ since all other quantities in the second term of (1) are known.

We suggest the following applications.

(a) Measurement of γ_n and thus of the magnetic moment of the neutron by measuring the scattering cross section for very slow neutrons or under small angles, so that $q \cong 0$ and $F(\mathbf{q})$ becomes practically one.

(b) Production of polarized neutron beams by letting neutrons pass through magnetized iron and observing that due to the cross product in the expansion of (1) the weakening of the beam due to scattering is different for neutrons with opposite orientation of their magnetic moment with respect to the magnetizing field. For example, the intensity of a neutron beam after having passed through two plates of iron should be different whether both are magnetized parallel to the beam or one is magnetized parallel and the other antiparallel.

(c) An experimental study of the distribution of the magnetizing electrons in ferromagnets, particularly whether they are conduction electrons or belong to inner shells, by investigating the scattering for different values of q and thus obtaining some information about the function $F(q)$ or the magnetic distribution function $g(q)$ related to it by Formula (2).

Experiments are under way here to test the predicted effect and its implications. Even if no magnetic scattering could be observed this should lead to the interesting conclusion that the magnetic moment of the neutron is considerably less than that theoretically to be expected.

Stanford University, July 6, 1936.

18 On the Magnetic Scattering of Neutrons*

O. Halpern and M. H. Johnson†

In this paper there is contained a full elaboration of two previously published short notes on the subject of magnetic scattering of neutrons together with a comprehensive treatment of certain sides of this problem which have already received some attention from other authors. After presenting the state of the problem in the introduction and discussing in detail our reasons for the choice of an interaction function between neutrons and electrons, and the nonmagnetic interaction between neutrons and nuclei, the various possible cases of coherent and incoherent scattering and depolarization phenomena are treated. Later applications to the theory of ferromagnetic scattering are kept in mind. The general expression for the cross section due to magnetic interaction is obtained and applied to various classes of phenomena (scattering by free, rigidly aligned, and coupled magnetic ions). The influence of the elastic form-factor is treated quantitatively with the aid of a simple model for the current distribution in the ion. Finally a series of performed or suggested experiments is discussed mainly from the point of view whether they will permit theoretical interpretation. Arrangements are described which will allow one to obtain a reliable value for the neutron's magnetic moment and also give insight into the magnetic constitution of the scatterer (ion or crystal) which will exceed the knowledge obtainable from macroscopic magnetic experiments.

Experiments

The considerations of the last sections obviously suggest a number of experiments in the field. The purpose of such experiments can essentially be twofold. They can be arranged with the aim of obtaining information about the magnetic moment

*Physical Review, 55, 898 (1939): only the abstract and final section of the paper are reproduced.

†New York University, University Heights, New York, New York. Received December 3, 1938.

of the neutron, or they can be used to explore the magnetic structure of the scattering system provided that the magnetic moment of the neutron is known. Quantitative information about it therefore seems to be of paramount importance.

The determination of the magnetic moment of the neutron can be carried out by scattering experiments alone or by a combination of scattering experiments with outside fields which act upon the polarization state of neutron beams. The last type of experiment is the one carried out by Frisch, Von Halban and Koch.[1] In these experiments a beam of neutrons was partially polarized by allowing it to pass through a magnetized piece of iron. Upon leaving the ferromagnet the neutrons were exposed to a magnetic field which could be varied in magnitude and direction, and which changed the state of polarization by an amount dependent on the neutron's magnetic moment during the passage from the first ferromagnet to a second. To determine the angle through which the individual spins rotate it is necessary to know the time of passage, i.e., the velocity of the neutrons. The scattering by, or the transmission through, the second ferromagnet will show a maximum (minimum) for definite directions of polarization of the incident beam. It is obvious that by assuming a certain knowledge about the velocity of the neutron, the variation of the strength of the magnetic field will permit a change from maximum to minimum transmission and thereby obtain information about the magnetic moment of the neutron. The authors gave as the most probable value two nuclear magnetons for the magnitude of the magnetic moment of the neutron.

Unfortunately the effect observed is very small, though, according to the authors, beyond the experimental error. It seems that iron is as poor a polarizer as it is an analyzer since the change in scattering (transmission intensity) remains a fraction of a percent. Furthermore, we do not seem to be able to determine the neutron velocity with too great an accuracy which also has a detrimental influence on the quantitative reliability of the method described.

---

[1]Frisch, Von Halban and Koch, *Phys. Rev.* **53**, 719 (1938).

Turning now to pure scattering and transmission experiments, it may, perhaps, be pointed out that the treatment of the scattering of a neutron by a *free* paramagnetic ion involves no uncertainties with the possible exception that the interaction between neutron and electron may not be wholly magnetic in origin. Furthermore, the scattering effects which are to be expected if the magnetic moment of the neutron is of the order of magnitude of one nuclear magneton can be very large (cf. (5.5)). Therefore, as previously stated, the simplest and most direct method of determining the neutron's moment seems, in our opinion, to be a scattering experiment with an appropriate paramagnetic salt. The salt selected should have a large magnetic moment, a condition which is well satisfied by the salts of divalent manganese and trivalent iron, and to a lesser extent by salts of several other ions in the iron group. (The rare earths are not very suitable for experiments of that type since they cannot be freed from highly absorbing members of their group). A particular salt must then be chosen which approximates as closely as possible the condition of a free ion. The ion will be called free if there is no appreciable effect from quenched orbital currents, from the spin-orbit coupling and from the spin-spin coupling between different ions. There is convincing evidence that divalent manganese and trivalent iron ions are in an S state so that the first two conditions are well satisfied. If we finally choose a salt which shows the full molar magnetic susceptibility as calculated on the assumption of a free spin (T_c in (8.12) small), we may be fairly certain to have also avoided difficulties arising from spin-spin coupling. Several salts whose susceptibility has been studied over a wide range of temperature satisfy (8.11) with a constant T_c considerably less than $100°K$ (e.g. $MnSO_4$). After selection of the salts as described, scattering experiments with slow neutrons should be carried out under several angles up to a minimum deflection chosen as small as the experimental arrangement will permit. These scattering measurements should be made relative to some standard which is nonmagnetic and shows only isotropic nuclear scattering; thereby geometric uncertainties can be avoided. The scattering data can then be fitted to the curve corresponding to a form factor (cf. Section VII) by an appropriate choice of the size of the scattering domain. Even without such theoretical assistance the measurement of the scattering at several angles should permit a satisfactory extrapolation to very small forward angles where the form factor becomes unity. The experiments can then be evaluated by the use of (5.9).

The various formulae of Section VI offer in principle a basis for the determination of the magnetic moment of the neutron through experiments with ferromagnets. By far the simplest method can be deduced from (6.40). It consists in an observation of the beam scattered by a magnetized body under a constant angle of scattering θ and at two different azimuths so chosen that $e\cdot\kappa = 1$ and $e\cdot\kappa = 0$. The difference in scattering then amounts to the *total magnetic scattering* independent of the amplitude and phase of the nuclear scattering. The disadvantages of this arrangement are several: It is necessary to delimit the scattered beam to a fraction of the azimuthal circumference which weakens the intensity by a factor of 2π approximately. It is furthermore essential to have single scattering since otherwise additional neutrons will be scattered into the chosen angle and also the assumption of a primarily unpolarized beam would no longer be justified. Apart from these not unsurmountable technical difficulties there remain certain theoretical ambiguities of perhaps greater importance. It would of course also be necessary, as in the paramagnetic case discussed before, to carry out observations at different angles θ to obtain information about the form factor. Even for the case of elastic scattering the form factor for ferromagnetic bodies will probably be of greater importance than in the case of paramagnetic scattering from free ions since it is likely that the outer shells will also be somewhat coupled and that there has occurred a non-calculable but probably large amount of spin quenching. A considerable contribution from incoherent magnetic scattering must also be expected to be present which will probably strongly diminish the otherwise marked azimuthal effect.

It should also be mentioned that the relative effect, even in the absence of all the difficulties mentioned above, is rather unfavorable in the case of iron, since the nuclear cross section is large, $d\Phi_n = 10^{-24}d\Omega$, while the elastic magnetic scattering amounts to $d\Phi_n \sim 2 \cdot 5 \times 10^{-25}d\Omega$ if we use for γ the empirically determined magnetic moment ~ 2 Bohr magnetons.

Originally it had been suggested[2] and an attempt[2] made to determine the magnetic moment with the aid of polarization experiments based upon the interference between nuclear and magnetic scattering. The two types of experiments mostly discussed are the double scattering and transmission arrangements which we mentioned in Section VI. Here too there exists in our opinion grave difficulties if one attempts to evaluate the experiments quantitatively for the purpose of determining the neutron moment. Section III dealt with the problem how far the coherent scattering of a nucleus can be determined from observation of the total nuclear cross section. We found that in the presence of isotopes and the nuclear spin it is almost impossible to draw quantitatively reliable conclusions. It must be admitted that the case of iron is favorable insofar as the dominant isotope has an atomic weight of 56 and therefore, probably no spin. It nevertheless does not seem feasible to determine the amplitude of coherent scattering quantitatively; we must be satisfied to enclose it within certain not improbable limits. There must, in addition, be considered a background due to inelastic nuclear scattering which arises from the coupling with the lattice. Since this coupling need not be the same for nuclear and magnetic scattering there arise inaccuracies which cannot be removed at our present state of knowledge. The inelastic scattering which is due to the strong spin coupling forces will, in the case of a saturated ferromagnet, also shown polarization effects of an unpredictable magnitude. The difficulty of obtaining sufficient intensity in a neutron beam for the purpose of carrying out a double scattering experiment has already been pointed out in the literature.

---

[2] Cf. e.g. Hoffman, Livingstone and Bethe, *Phys. Rev.* **51**, 214 (1937); several notes by J. Dunning and collaborators in *Phys. Rev.* **51** (1937).

That all these factors mentioned are of considerable influence on the actual state of polarization of a beam passing through iron follows in our opinion from the minuteness of the effect which has been observed in the experiments by Frisch, von Halban and Koch. A rough estimate on the basis of (6.30) etc. would indicate that iron is by far a better polarizer (and analyzer) than the actual observations indicate. We consider these experiments as offering direct support for our contention as to the complexity of scattering processes in ferromagnetic bodies.

Similar difficulties enter into the quantitative evaluation of experiments concerned with the attenuation of an incident unpolarized neutron beam. Since it is the *total* and not the *differential* cross section that here becomes of importance, we are not able to study independently the form factor as function of the scattering angle. This form factor enters (cf. e.g. (6.53)) in a decisive but complex manner into the absorption cross section; without knowing the form factor no conclusions can be drawn as to the magnitude of the magnetic moment. If there is an appreciable amount of incoherent nuclear scattering present, the beam will be depolarized as it passes through the ferromagnet. Care must also be taken that in the case of a highly attenuated primary beam no secondary neutrons shall be scattered into the forward direction. Previous evaluations have also neglected the purely magnetic coherent and incoherent scattering as well as the effect of the precession of the neutron spin during the passage through the ferromagnet.[3] The combined weight of all these variable factors seems, in our opinion, sufficient to make a quantitative evaluation of transmission experiments appear rather improbable.

Assuming a quantitative knowledge of the magnetic moment

---

[3]*Note added in proof.*—The problem of magnetic scattering of neutrons has been further treated together with related questions in two notes which will appear shortly. Abstracts have been presented to the meetings of the American Physical Society in Washington (cf. a paper by O. Halpern and Th. Holstein presented in December 1938; *Phys. Rev.* **55**, 601 (1939); and a paper by Halpern, Hammermesh and Johnson presented in April 1939; Abstract No. 73 in *Am. Phys. Soc. Bull.* **14**, No. 2, April 12, 1939).

of the neutron which, as we have shown, can probably be obtained with fewest ambiguities from observations of paramagnetic scattering, we are able to outline a series of experiments from which much information can be gained as to the magnetic structure of the scatterer. As previously pointed out, we find ourselves now on less satisfactory ground as far as the theory is concerned, and we think that the most promising attack can be made from the experimental side. The difficulties to which we are referring are mainly our lack of knowledge concerning the elastic and even more important, the inelastic form factors. The complications which we encounter in attempting to estimate the energy change per collision due to the presence of spin coupling forces have been discussed fully in Section VIII. We there realized the impossibility of drawing reliable conclusions from the mean square fluctuation of the energy which, in our opinion, does not at all determine the actual energy change per collision. Ceding this group of problems to experimental investigation, we have learned in the meantime by letter and personal communication that Professor Van Vleck has become interested in the theoretical side of this problem and is attempting to derive information about the energy changes from considerations about energy fluctuations. His results show that from such considerations the average energy change is of the order of the exchange integral.

Leaving these theoretical considerations aside for the moment we may point out that there exists a very simple type of experiment which will show if it is justified to predict energy changes of the order of magnitude of the exchange integral J. For this purpose it is only necessary to determine the integral cross section of various salts containing the same paramagnetic ion (Mn^{++} or Fe^{+++}). Since these salts show widely varying susceptibilities and exchange integrals (Curie temperatures between $100°$ and $1000°K$), we should expect large changes in the integral cross section. If, for example, we choose a salt like MnS which does not show an appreciable magnetic scattering in the backward direction, then we should expect according to the hypothesis mentioned a vanishingly small total magnetic cross section, since the inelastic form factor will almost completely eliminate the forward scattering also. If, on the other hand, the small backward scattering is due to the influence of the elastic form factor which (cf. (7.30)) is small enough to make it unobservable, then a sizable magnetic cross section should be measured which would be due to a large amount of forward scattering. All these discussions presuppose of course, the existence of a magnetic moment of the neutron which is not much smaller than one nuclear magneton.

Experimenting with quasi-free magnetic ions we can use the angular dependence of the magnetic scattering to obtain information on the magnetically active size of the ion. Similarly observations on the angular dependence of scattering can be used to obtain information about the probabliity $P(L, L')$. For this purpose it seems advisable to carry out observations at various neutron energies, and for a series of salts of variable "magnetic dilution". Since the magnitude of spin coupling is known from susceptibility data, one can thus obtain experimentally a relationship between spin coupling and neutron scattering. Again the salts of divalent manganese and trivalent iron are especially suitable because of the absence of orbital currents. It will also prove interesting to investigate the scattering from paramagnetic metals. Susceptibility data are here difficult to interpret because of the erratic temperature dependence which may well be connected with a large spin coupling in these magnetically concentrated materials. This spin coupling will in the antiferromagnetic case operate to quench the ionic spins in the presence of an external magnetic field. After one has determined the general effect of spin coupling with the aid of observations on ions of known magnetic structure, the neutron scattering from metals should present information as to the actual ionic moment and hence, the ionic state in the metal.

We have already pointed out that the presence of spin coupling forces could give rise to marked phenomena if the scattering by ferromagnets is observed above and below the Curie point. If the energy changes should really turn out to be of the order of magnitude of the exchange integral, then we would expect that the scattering above the Curie point is practically absent on account of the large inelastic form factor. An experiment of such a kind should give us information similar to that obtained from observations on the total cross sections of paramagnetic salts.

19 Detection of Antiferromagnetism by Neutron Diffraction*

C. G. SHULL† and J. SAMUEL SMART‡

Two necessary conditions for the existence of ferromagnetism are: (1) the atoms must have a net magnetic moment due to an unfilled electron shell, and (2) the exchange integral J relating to the exchange of electrons between neighboring atoms must be positive. This last condition is required in order that spin states of high multiplicity, which favor ferromagnetism, have the lowest energy. It seems certain that for many of the non-ferromagnetic substances containing a high concentration of magnetic atoms the exchange integrals are negative. In such cases the lowest energy state is the one in which the maximum number of anti-parallel pairs occur. An approximate theory of such substances has been developed by Néel,[1] Bitter,[2] and Van Vleck[3] for one specific case and the results are briefly described below.

Consider a crystalline structure which can be divided into two interpenetrating lattices such that atoms on one lattice have nearest neighbors only on the other lattice. Examples are simple

*Physical Review, 76, 1256–7 (1949). This work was supported in part by the ONR.

†Oak Ridge National Laboratory, Oak Ridge, Tennessee.

‡Naval Ordnance Laboratory, White Oak, Silver Spring, Maryland. August 29, 1949.

[1] L. Néel, Ann. de physique 17, 5 (1932).
[2] F. Bitter, Phys. Rev. 54, 79 (1938).
[3] J. H. Van Vleck, J. Chem. Phys. 9, 85 (1941).

cubic and body-centered cubic structures. Let the exchange integral for nearest neighbors be negative and consider only nearest neighbor interactions. Theory then predicts that the structure will exhibit a Curie temperature. Below the Curie temperature the spontaneous magnetization *vs.* temperature curve for one of the sub-lattices is that for an ordinary ferromagnetic material. However, the magnetization directions for the two lattices are antiparallel so that no net spontaneous magnetization exists. At absolute zero all of the atoms on one lattice have their electronic magnetic moments aligned in the same direction and all of the atoms on the other lattice have their moments antiparallel to the first. Above the Curie temperature the thermal energy is sufficient to overcome the tendency of the atoms to lock antiparallel and the behavior is that of a normal paramagnetic substance.

Materials exhibiting the characteristics described above have been designated "antiferromagnetic". Up to the present time the only methods of detecting antiferromagnetism experimentally have been indirect, e.g., determination of Curie points by susceptibility and specific heat anomalies. It has occurred to one of us (J.S.S.) that neutron diffraction experiments might provide a direct means of detecting antiferromagnetism. In an antiferromagnetic material below the Curie temperature a rigid lattice of magnetic ions is formed and the interaction of the neutron magnetic moment with this lattice should result in measurable coherent scattering. Halpern and Johnson[4] have shown that the magnetic and nuclear scattering amplitudes of a paramagnetic atom should be of the same order of magnitude and this result has been qualitatively verified by experimental investigators.[5] At the time of the above suggestion, an experimental program on the determination of the magnetic scattering patterns for various paramagnetic substances (MnO, MnF_2, $MnSO_4$ and Fe_2O_3) was

---

[4]O. Halpern and M. H. Johnson, *Phys. Rev.* **55,** 898 (1939).

[5]Whittaker, Beyer, and Dunning, *Phys. Rev.* **54,** 771 (1938); Ruderman, Havens, Taylor, and Rainwater, *ibid.* **75,** 895 (1949); and also unpublished work at Oak Ridge National Laboratory.

underway at Oak Ridge National Laboratory and room temperature examination had shown (1) a form factor type of diffusion magnetic scattering (no coupling of the atomic moments) to exist for MnF_2 and $MnSO_4$, (2) a liquid type of magnetic scattering (short-range order coupling of oppositely directed magnetic moments) to exist for MnO and (3) the presence of strong coherent magnetic diffraction peaks at forbidden reflection

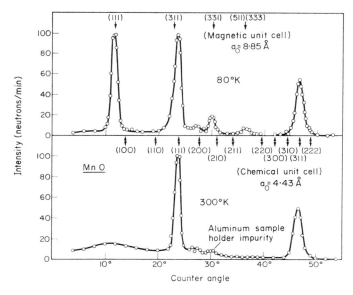

FIG. 1. Neutron diffraction patterns for MnO at room temperature and at 80°K.

positions for the α-Fe_2O_3 lattice. The latter two observations are in complete accord with the antiferromagnetic notion since the Curie points for MnO and α-Fe_2O_3 are respectively[6] 122°K and 950°K.

---

[6]Bizette, Squire, and Tsai, *Comptes Rendus* **207**, 449 (1938).

Figure 1 shows the neutron diffraction patterns obtained for powdered MnO at room temperature and at 80°K. The room temperature pattern shows coherent nuclear diffraction peaks at the regular face-centred cubic reflection positions and the liquid type of diffuse magnetic scattering in the background. It should be pointed out that the coherent nuclear scattering amplitudes for Mn and O are of opposite sign so that the diffraction pattern is a reversed NaCl type of pattern. The low temperature pattern also shows the same nuclear diffraction peaks, since there is no crystallographic transition in this temperature region,[7] and in addition shows the presence of strong magnetic reflections at positions not allowed on the basis of the chemical unit cell. The magnetic reflections can be indexed, however, making use of a magnetic unit cell twice as large as the chemical unit cell. A complete description of the magnetic structure will be given at a later date.

In conclusion it appears that neutron diffraction studies of antiferromagnetic materials should provide a new and important method of investigating the exchange coupling of magnetic ions.

[7]B. Ruhemann, *Physik. Zeits. Sowjetunion* **7**, 590 (1935).

20 Neutron Scattering and Polarization by Ferromagnetic Materials*

C. G. SHULL, E. O. WOLLAN, and W. C. KOEHLER†

Neutron diffraction studies are reported on a series of magnetized and unmagnetized ferromagnetic materials. The diffraction patterns for unmagnetized, polycrystalline samples of Fe and Co are found to possess both nuclear and magnetic components with the latter in agreement with the magnetic scattering theory with respect both to intensity of scattering and form factor angular variation. Studies on the magnetic structure of Fe_3O_4 are shown to strongly support Néel's proposed ferrimagnetic structure. Predictions of the theory regarding intensity effects upon sample magnetization are fully confirmed and the Schwinger-Halpern-Johnson formulation of the interaction function between the neutron's magnetic moment and the internal fields in a ferromagnet is substantiated. A pronounced variation of intensity around the Debye ring in the diffraction pattern for a magnetized sample is found. Neutron polarization effects in the Bragg scattered beams from magnetized crystals of Fe and Fe_3O_4 have been studied and it is shown that very highly polarized beams are obtained for certain reflections. This method of monochromatic beam polarization is found to compare very favorably with other methods with respect to polarization value, beam intensity, and ease of obtainment.

Introduction

The theory of the scattering and polarization of neutrons by ferromagnetic substances has been given a very general treatment by Halpern and co-workers.[1] Up to the present time, experi-

*Physical Review, **84,** 912–21 (1951).

†Oak Ridge National Laboratory, Oak Ridge, Tennessee. Received August 20, 1951.

[1] O. Halpern and M. H. Johnson, *Phys. Rev.* **55,** 898 (1939). O. Halpern and T. Holstein, *ibid.* **59,** 960 (1941). Halpern, Hamermesh, and Johnson, *ibid.* **59,** 981 (1941).

mentation in this field has centred upon studies of the single and double transmission effect[2] and more recently upon studies of critical reflection from magnetized mirrors.[3]

The present report deals with studies of the intensity distribution and the polarization of the scattered neutron radiation from both unmagnetized and magnetized ferromagnetic substances. These studies give information on the form factor dependence of magnetic scattering, on the basic nature of the neutron's magnetic interaction and on the magnetic structure existing in certain ferromagnetics, *viz.*, the spatial distribution of the various magnetic ions within the ferromagnetic lattice.   A previous report[4] has discussed neutron scattering by paramagnetic and antiferromagnetic lattices and brief reports of some aspects of the present work concerning the neutron's magnetic interaction and the polarization phenomena have been given in the literature.[5]

The general expression for the differential scattering cross section of a magnetic ion has been given by Halpern and Johnson[1] (H-J) as

$$F^2 = C^2 + 2CD\mathbf{q}\cdot\boldsymbol{\lambda} + D^2q^2 \qquad (1)$$

where C is the nuclear scattering amplitude, D the magnetic scattering amplitude, λ a unit vector describing the polarization state of the neutron being scattered,

$$\mathbf{q} = \mathbf{e}(\mathbf{e}\cdot\boldsymbol{\kappa}) - \boldsymbol{\kappa} \quad \text{and} \quad q^2 = 1 - (\mathbf{e}\cdot\boldsymbol{\kappa})^2 \qquad (2)$$

where $\mathbf{e}$ is the unit scattering vector and $\boldsymbol{\kappa}$ a unit vector parallel to the magnetic moment vector of the magnetic ion. The magnetic scattering amplitude D is given by

$$D = (e^2/mc^2)\gamma Sf = 0.539 \times 10^{-12} Sf \,\text{cm} \qquad (3)$$

[2]Hughes, Wallace, and Holtzman, *Phys. Rev.* **73**, 1277 (1948). Fleeman, Nicodemus, and Staub, *ibid.* **76**, 1774 (1949). Burgy, Hughes, Wallace, Heller, and Woolf. *ibid.* **80**, 953 (1950).

[3]D. J. Hughes and M. T. Burgy, *Phys. Rev.* **81**, 498 (1951).

[4]Shull, Strauser, and Wollan, *Phys. Rev.* **83**, 333 (1951).

[5]Shull, Wollan, and Strauser, *Phys. Rev.* **81**, 483 (1951). C. G. Shull, *ibid.* **81**, 626 (1951).

where γ is the neutron magnetic moment expressed in nuclear Bohr magnetons, S the spin quantum number of the magnetic atom, and f the magnetic form factor characteristic of the electrons responsible for the atomic magnetic moment.

For unpolarized incident neutron radiation, $\mathbf{q} \cdot \boldsymbol{\lambda}$ averages to zero and the differential cross section is the sum of the nuclear and magnetic contributions

$$F^2 = C^2 + D^2 q^2. \tag{4}$$

The numerical value for q^2 will depend according to Eq. (2) upon the relative directions of the scattering vector and the magnetization vector. A further discussion of the form of this dependence will be given in a later section of this paper.

Diffraction by unmagnetized ferromagnetic materials

We consider first the scattering by the ferromagnetic elements Fe and Co with no applied magnetic field. At temperatures below the Curie point, the magnetic moments of the atoms are in parallel alignment within a single domain but, without an applied field, the magnetic orientation from one domain to the next will, on the average, be random. Each domain constitutes a magnetic crystallite and since the magnetic moments are aligned within the domains, the magnetic scattering will appear in the Bragg peaks together with the nuclear scattering, the intensity of the two contributions being additive according to Eq. (4). The magnetic and nuclear contributions can be separately determined since the magnetic scattering will exhibit a form factor fall off with scattering angle due to the spatial distribution of the magnetic electrons within the atom while the nuclear scattering cross section is spherically symmetric. Following a discussion on the diffraction results for these ferromagnetic elements, data and their interpretation for a more complicated ferromagnetic substance, Fe_3O_4 (magnetite), will be given.

Iron. Figure 1 shows the powder diffraction pattern taken for a sample of polycrystalline Fe with neutrons of wavelength

1·204Å. Rather high intensities are obtained for this material primarily because of the large coherent nuclear scattering amplitude for Fe. A number of samples were studied, with most emphasis on samples of filings taken from pure Armco iron blocks. The filings were given a heat treatment and hydrogen reduction to remove any oxide layer and some of the cold-working effects. The samples were contained in flat cells of thickness about $\frac{1}{2}$ in. between thin quartz windows. Corrections for second-order

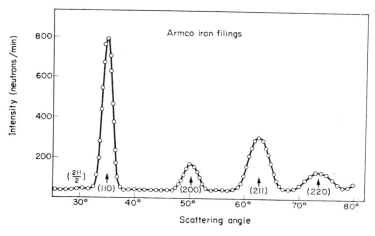

Fig. 1. Neutron diffraction pattern for polycrystalline Fe.

wavelength contribution were made to the patterns and aside from these low intensity second-order parasitic reflections, all of the observed reflections could be indexed on the basis of the usual body-centred cubic reflections for the normal cubic cell. The observed integrated intensities were placed on an absolute scale by comparison with a standard sample of nickel powder for which we have measured the coherent scattering cross section as 13·4 barns.

The integrated intensities shown in the pattern of Fig. 1 were converted to differential scattering cross sections in the usual

M

fashion[6] and these values for the four observed reflections are shown in Fig. 2. It is to be seen that this cross section shows a significant, although not pronounced, angular variation and this is because of the large isotropic nuclear scattering relative to the magnetic scattering. Using the magnetic moment per Fe atom of 2·22 Bohr magnetons as determined from saturation magnetization studies on iron, and hence a value of 1·11 for S_{eff} since the

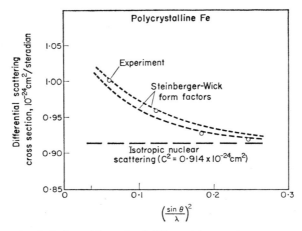

FIG. 2. Variation with angle of differential scattering cross section per Fe atom for the powder pattern reflections. The scattering cross section contains an angularly-dependent magnetic component and an isotropic nuclear component. The calculated magnetic scattering (with the Steinberger-Wick form factors) is shown.

gyromagnetic ratio for Fe is very close to 2, one evaluates with Eq. (3) the expected magnetic scattering amplitude.

$$D(\text{Fe}) = 0·598f \times 10^{-12} \text{ cm.} \tag{5}$$

Steinberger and Wick[7] have given the results of a recent calculation of the magnetic form factor f for iron and on the basis of these values, the best fit between the data and Eqs. (4) and (5)

[6]C. G. Shull and E. O. Wollan, *Phys. Rev.* **81,** 527 (1951).

[7]J. Steinberger and G. C. Wick, *Phys. Rev.* **76,** 994 (1949).

is shown on the graph of Fig. 2. Also shown on the figure are extremes in the Steinberger-Wick calculation for different assumptions about the asymptotic behavior of the 3d-wave functions. The experimental data are not considered accurate enough to distinguish between one selection or another. Previous studies by Hughes, Wallace, and Holtzman[2] on the single transmission effect (change in transmission upon magnetization) with monochromatic neutrons have also shown general agreement with the magnetic scattering theory and the Steinberger-Wick form factor evaluation.

In the foregoing evaluation of the magnetic scattering cross section D^2 from the experimental data, q^2 has been taken as $\frac{2}{3}$ which is its average value when the magnetic moment within a domain has equal probability of all orientations relative to the crystal axes. It is generally agreed, however, that in iron the moments have a high probability of being aligned along the cube edges which are the easiest directions of magnetization in a single crystal. It can be shown, however, that even in this case the average value of q^2 will be $\frac{2}{3}$ for all (hkl) reflections with a polycrystalline and polydomain ferromagnetic sample. This fact makes it evident that the present data cannot be used to indicate the direction relative to the crystal axes of domain magnetization. This is in contrast to the conclusions[4] which have been drawn from antiferromagnetic reflections in MnO and FeO at low temperature wherein it was possible to establish the moment orientations even with polycrystalline data because incomplete multiplicities were obtained with these particular magnetic structures.

In addition to the studies of polycrystalline iron described above, a series of samples of isotopically enriched iron was also studied. Preliminary studies to determine the nuclear scattering phase of Fe^{54}, Fe^{56}, and Fe^{57} were carried out with the enriched iron in the form of Fe_2O_3, after which the metals in a finely divided state were studied in greater detail. For each of the isotopic iron samples, as for normal iron, the differential scattering cross section exhibited an angular variation. In the case of Fe^{56} the residual magnetic scattering after subtraction of the nuclear contributions was identical, within experimental error, with that obtained from normal iron, as indeed, is to be expected. A direct determination of the magnetic scattering from Fe^{54} and Fe^{57}, for which the nuclear scattering amplitudes are more favorable, did not prove feasible because of the very small amounts of sample at our disposal. In these two latter instances, the magnetic contribution was assumed known and equal to that for normal iron and Fe^{56}, and the nuclear scattering amplitude was obtained by subtraction.   In Table I are shown the nuclear scattering

TABLE I. THE COHERENT NUCLEAR SCATTERING
CROSS SECTIONS OF IRON AND ITS ISOTOPES

Isotope	Coherent scattering amplitude (10^{-12} cm)	Coherent scattering cross section (barns)
Fe^{54}	$+0.42$	2.2 ± 0.1
Fe^{56}	$+1.00$	12.6 ± 0.2
Fe^{57}	$+0.23$	0.64 ± 0.05
Fe	$+0.96$	11.5 ± 0.2

properties of the isotopes of iron obtained in this way. The total scattering cross section of iron, using these isotopic cross sections, may be calculated and is found to be 11·6 barns. Comparison of this value with the coherent scattering cross section of 11·5 barns indicates that the isotopic diffuse scattering in normal iron is very small. This result supports the conclusions drawn by Hughes and Burgy[8] from their investigations of the single transmission effect in a single crystal of iron with neutron energies below that corresponding to the first Bragg peak in which they find less than 0·25 barn of spin dependent or isotopic diffuse scattering.

Cobalt. Diffraction patterns for a sample of face-centred-cubic cobalt in the form of metallic filings were obtained in the same fashion as in the above study on iron. The resulting scattering

8D. J. Hughes and M. T. Burgy, *Phys. Rev.* **80**, 481 (1950).

cross sections per magnetic atom are shown in Fig. 3 with the data resolved into its nuclear and magnetic contributions. Because of the much smaller nuclear scattering here than was found for iron, the magnetic scattering effects show up more strikingly.

Using the experimentally determined magnetic moment for Co of 1·74 Bohr magnetons[9] from magnetization studies on cubic cobalt, the magnetic scattering amplitude is evaluated as

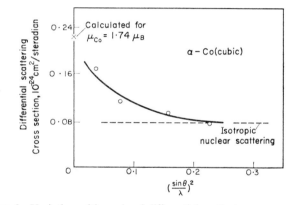

FIG. 3. Variation with angle of differential scattering cross section per Co atom for the Co powder pattern reflections.

$0.468f \times 10^{-12}$ cm. The value of the cross section corresponding to this amplitude at zero scattering angle (unit form factor) is shown on the ordinate axis of Fig. 4. The form factor describing the magnetic scattering is that suggested by the data and it is seen that this extrapolates satisfactorily to the calculated forward-direction cross section. It would be expected that this form factor would be closely similar to that calculated for iron by Steinberger and Wick and indeed this is substantiated in a quantitative

[9] A. J. P. Meyer and P. Taglang, *Compt. rend.* **231,** 612 (1950).

comparison of the two. The slightly larger nuclear charge for Co would compress the 3d magnetic shell somewhat and the form factor would be slightly less angularly dependent than would be the case for iron. There is a suggestion of this in the Co experimental data.

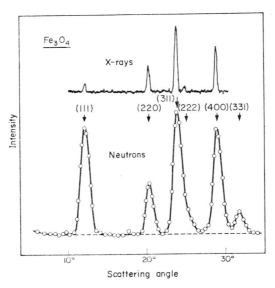

FIG. 4. Comparison between X-ray and neutron diffraction patterns for Fe_3O_4 (magnetite) at room temperature. Pronounced magnetic scattering contribution in the neutron pattern is found.

As seen from the graph the coherent nuclear scattering cross section for Co amounts to only 1·0 barn and this is a consequence of the pronounced nuclear spin incoherent scattering for this nucleus. This has been discussed in an earlier report.[6] The magnetic scattering is comparable to or exceeds the nuclear scattering in the forward direction and this fact leads to interesting polarization phenomena in the neutron scattering by magnetized samples as will be discussed in a later section.

Fe_3O_4 (*magnetite*). This oxide possesses a spinel-type structure and is ferromagnetic on a macroscopic scale to about the same extent as nickel. There are both divalent and trivalent Fe ions present in the structure and the molecular formulation can be considered as containing one Fe^{++} ion and two Fe^{+++} ions. Verwey and Heilmann[10] consider the structure to be that of an inverted spinel in which the octahedral Fe sites are occupied at random by equal numbers of divalent and trivalent ions whereas the tetrahedral sites are occupied only by trivalent ions. Further considerations by Néel[11] on the magnetic properties suggest a ferrimagnetic structure for this material in which the magnetic moments of the tetrahedral ions are coupled antiferro-magnetically to those of the octahedral ions. Because the latter are in the majority, Fe_3O_4 will be resultantly ferromagnetic and Néel has been able to explain quantitatively the observed magnetic moment per Fe atom obtained from magnetization data.

Since neutron scattering by such a magnetic lattice will be sensitive to the strength, position, and relative orientation of the magnetic ions, the neutron diffraction pattern should offer direct evidence as to the validity of these magnetic structure suppositions. Several powder samples have been examined, the principal quantitative measurements being made on a very pure sample kindly supplied by Professor A. von Hippel. A typical diffraction pattern taken at room temperature is shown in Fig. 4. As in the case of the simple ferromagnetic elements already discussed, there will be no coherence between the nuclear and magnetic portions of the pattern. Since the nuclear scattering amplitudes for iron and oxygen are known accurately and the crystallographic structure has been well established from X-ray studies, the nuclear scattering contribution in the neutron pattern can be reliably accounted for. The remaining portion can be attributed to magnetic scattering and consequently can be compared to that predicted by any suggested magnetic structure.

[10]E. J. W. Verwey and E. L. Heilmann, *J. Chem. Phys.* **15**, 174 (1947). Verwey, Haayman, and Romeijn, *ibid.* **15**, 181 (1947).
[11]L. Néel, *Ann. phys.* **3**, 137 (1948).

The comparison between the observed and the calculated intensities for the Néel structural model is given in Table II along with the crystal structure factors which characterize the various reflections. In this tabulation, f_1, f_2, and f_0 represent the scattering amplitudes of a tetrahedral iron ion, an octahedral iron ion, and an oxygen ion respectively. For the nuclear scattering

TABLE II. COMPARISON BETWEEN OBSERVED INTENSITIES FOR Fe_3O_4 AND THOSE CALCULATED FOR THE NÉEL MAGNETIC STRUCTURE MODEL

Reflection	Crystal structure factor	Calculated intensities			Observed intensity
		Magnetic	Nuclear	Total	
(111)	$2(4f_1 - 4\sqrt{2}f_2)^2$	902	32	934	860
(220)	$(8f_1)^2$	125	218	343	360
(311)	$2(4f_1 + 4\sqrt{2}f_2)^2$ $\left.\begin{array}{c}\\\\\end{array}\right\}$	112	948	1060	1070
(222)	$(-16f_2 + 32f_0)^2$				
(400)	$(8f_1 - 16f_2 - 32f_0)^2$	116	649	765	780
(331)	$2(4f_1 - 4\sqrt{2}f_2)^2$	94	16	110	135
(422)	$(8f_1)^2$				
(333)	$2(4f_1 + 4\sqrt{2}f_2)^2$ $\left.\begin{array}{c}\\\\\end{array}\right\}$	20	670	690	700
(511)	$2(4f_1 + 4\sqrt{2}f_2)^2$				
(440)	$(8f_1 + 16f_2 + 32f_0)^2$ $\left.\begin{array}{c}\\\\\end{array}\right\}$	32	1692	1724	1730
(531)	$2(4f_1 - 4\sqrt{2}f_2)^2$				

All intensities (neutrons/min) are on an absolute scale.

amplitudes, that for iron is taken as $0.956 \cdot 10^{-12}$ cm and for oxygen $0.575 \cdot 10^{-12}$ cm. On the Néel magnetic model,

$$(f_1)_{mag} = -D(Fe^{+++})$$

and

$$(f_2)_{mag} = +\tfrac{1}{2}D(Fe^{+++}) + \tfrac{1}{2}D(Fe^{++}), \tag{6}$$

with reversed signs for the two amplitudes because of the postulated antiferromagnetic orientation of tetrahedral and octahedral ions. The spin quantum numbers for Fe^{+++} and Fe^{++} ions are, respectively, 5/2 and 2 and hence the appropriate magnetic scattering amplitudes are calculable from Eq. (3) as

$$D(Fe^{++}) = 1 \cdot 088f \times 10^{-12} \text{ cm}, \tag{7}$$

$$D(Fe^{+++}) = 1 \cdot 360f \times 10^{-12} \text{ cm}.$$

Finally, the total intensity in an (hkl) reflection, say the (111) reflection, is taken as

$$I_{111} = I_{\text{nuclear}} + I_{\text{magnetic}} = k\{2[4f_{\text{Fe}} - 4\sqrt{2}f_{\text{Fe}}]^2\}_{\text{nucl}}$$
$$+ \tfrac{2}{3}k\{2[4f_1 - 4\sqrt{2}f_2]^2\}_{\text{mag}} \tag{8}$$

in which k is an instrumental parameter determinable from the powder diffraction formula. In the numerical evaluation, the magnetic form factor for Mn^{++} (reference 4) has been used for the Fe ions since the latter have not been established accurately at the present time. This should be a good approximation since the Mn ion and Fe ion would not be expected to differ much in their 3d-shell characteristics.

The intensities shown in Table II are all on an absolute scale and the agreement between calculated and observed total intensity values appears satisfactory. Magnetized sample studies to be discussed in the next section permit a direct resolution of the total intensity into the magnetic and nuclear components and these also support the validity of the Néel structural model. Other magnetic models have been investigated and it is found that the calculated intensity is rather sensitive to the selected model. For instance, an alternative structure would be for the Fe^{++} ions to exist wholly at tetrahedral sites and the Fe^{+++} only at octahedral sites (a normal spinel structure). The observed magnetization could be obtained for this model by having the Fe^{++} ions coupled ferromagnetically but with no orientation coupling between the Fe^{+++} ions themselves or between Fe^{+++} and Fe^{++}. Thus the trivalent ions would be in a paramagnetic state. Intensity calculations for this model are in violent disagreement with the data, with the (111) intensity for instance, being eight times lower than observed. Thus the Néel model of anti-ferromagnetic coupling between the tetrahedral ions and the octahedral ions in the inverted spinel structure appears fully substantiated. Figure 5 shows a portion of the magnetic unit cell of Fe_3O_4 with this antiferromagnetic arrangement of the ions.

Some diffraction data were also obtained for Fe_3O_4 at low temperatures (80°K) to see if any structural changes developed

Fe_3O_4 – spinel structure

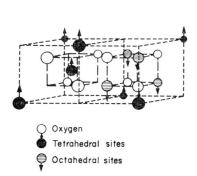

○ Oxygen
● Tetrahedral sites
⊖ Octahedral sites

FIG. 5. Portion of the magnetic unit cell for Fe_3O_4. The octahedral and tetrahedral Fe ions are coupled antiferromagnetically.

upon cooling. Fe_3O_4 exhibits interesting changes in magnetic properties and electrical conductivity at temperatures of the order 120–150°K and Verwey, Haayman, and Romeijn[10] have suggested that the pronounced increase in resistivity in lowering the temperature through 120°K is to be accounted for by an ordering of the Fe^{++} and Fe^{+++} ions at the octahedral sites. On this picture the relatively low resistivity at room temperature is caused by ionic migration (electron exchange) among the randomly arranged ions at the octahedral sites. The neutron diffraction patterns at high and low temperatures were found to possess no significant difference (within about 1 percent) and a detailed examination of the crystal structure factors and the expected intensities for the Verwey model showed that this was not inconsistent with the model. The redistribution of the Fe^{++} and Fe^{+++} ions, which do not differ too much in magnetic scattering power (see Eq. (7)) does not appear to affect the powder intensities appreciably. If studies were to be made on single crystal, single domain samples where one does not have the superposition of various (*hkl*) reflections, then indeed the ordering effect should be noticeable but this has not been performed experimentally. The powder data do show, however, that the basic antiferromagnetic structure is maintained at low temperatures since any departure from ferromagnetic coupling between various octahedral ions, with consequent change of sign in the scattering amplitude, affects the calculated intensity pronouncedly.

Scattering by magnetized samples

The magnetic contribution to the observed intensity of scattering in the diffraction pattern will depend according to Eq. (2) upon the relative orientation of the scattering vector and the ionic magnetic moment vector. Since the latter can be controlled in a ferromagnetic sample by application of a suitably high external magnetic field, there should be intensity changes when the field is altered either in strength or direction of application. By comparing the observed change with that calculated from the theory, it is possible to determine directly the magnetic component of the observed intensity and this would serve as an independent test on the magnetic scattering aside from the form-factor analysis discussed in the last section.

The diagrams shown in Fig. 6 illustrate the experimental arrangements of interest here. A monochromatic beam of neutrons is incident upon the polycrystalline ferromagnetic sample which is located within the pole gap of an electromagnet. The intensity of scattering in a portion of one of the Debye rings produced by the sample is studied (1) for the unmagnetized sample, (2) with the sample magnetized in a direction parallel to the scattering vector (arrangement (a) in the figure), and (3) with the sample magnetization perpendicular to the scattering vector [arrangement (b)]. According to Eqs. (2) and (4) the magnetic cross sections and hence intensities should have relative values for these three cases, $\frac{2}{3}$, 0, and 1, respectively.

For this purpose an electromagnet capable of producing a magnetic field of about 8000 oersteds in a 1-in. gap was mounted

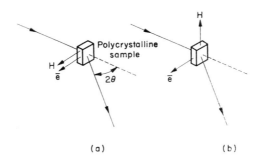

FIG. 6. Diagram of experimental arrangement used in the study of neutron scattering by magnetized, ferromagnetic substances. In (a) the applied magnetic field is in the plane of scattering and parallel to the scattering vector. In (b) the field is perpendicular to both.

on the axis of the conventional powder diffraction spectrometer. The orientation of this field relative to the plane of scattering could be selected by suitable arrangement of the magnet. The state of magnetization of the sample was followed by search coil measurements of the magnetic field intensity at the surface of the sample. Normal sample size was about 1 in. × $1\frac{1}{2}$ in. in area and thickness $\frac{1}{4}$ in. or $\frac{3}{8}$ in. with the field application along the 1 in. direction.

The first measurements of this type were made with small blocks of pure Armco iron with studies of the (110) and (200) intensity as a function of magnetization strength and direction. It was found that the intensity of scattering was decreased relative to the unmagnetized sample value when the field was applied parallel to the scattering vector, as is expected from the theory. On the other hand, the intensity also decreased when the sample was magnetized perpendicular to the scattering vector, an effect which was not at first expected. The decrease in this case arises from the fact that in the scattering process for the case of field perpendicular to the scattering vector the neutrons are partially polarized and these will experience a different transmission cross section in their passage through the polycrystalline iron block. When the attenuation effects characteristic of the partially polarized beam produced upon scattering are properly taken into account, one calculates for the (110) Fe reflection a decrease upon magnetization of 4·8 percent which agrees quite well with the experimentally observed value of 5·1 percent. Thus in spite of the larger cross section for scattering when $q^2 = 1$, the scattered beam is attenuated more in the sample and the expected intensity change is reversed.

The intensity effects which arise with iron upon magnetization are quite small as illustrated and this is because the magnetic scattering amplitudes are small in comparison to the unfavorably large nuclear scattering amplitude. More favorable cases are suggested for certain reflections of Co and Fe_3O_4 already discussed. The nuclear scattering amplitude for Co is quite small and should permit very large intensity effects upon magnetization but it is difficult to saturate magnetically. Magnetite on the other hand is rather easy to saturate magnetically and some of its reflections are very favorable for the present purpose. The (111) Fe_3O_4 reflection is almost completely magnetic in origin (see Table II) and this offers two advantages: (a) the intensity will be strongly affected by magnetization strength and direction and (b) there is very little polarization produced upon scattering, unlike the iron and cobalt cases, so that anomalous transmission effects will not be encountered. Because of these interesting and favorable properties, magnetite was studied in some detail.

Compressed powder samples of Fe_3O_4 were held between the pole pieces of the electromagnet and again field strength measure-

ments were made with a search coil at the surface of the sample.
Figure 7 shows typical traversals taken over the (111) Fe_3O_4
powder reflection with the sample block unmagnetized and when
magnetized at high field strength in the two directions shown
schematically in Fig. 6. It is seen that the scattered intensity
varies pronouncedly with field direction. Figure 8 summarizes
the data taken with varying field strength for the two field

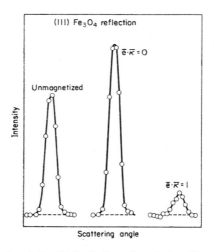

FIG. 7. Study of the (111) Fe_3O_4 reflected for different magnetization directions. The intensity is seen to depend strongly on the
sample magnetization direction with respect to the scattering vector.

orientations and shows the saturation effects characteristic of
increasing domain alignment with increasing field strength.

On Fig. 8 are shown the cross sections, $(C^2 + \frac{2}{3}D^2)$, C^2, and
$(C^2 + D^2)$ which correspond to the unmagnetized and two
magnetized states. The nuclear cross section term C^2 is very much
smaller than D^2 for this particular Fe_3O_4 reflection, in good
agreement with the structure factor calculations outlined in an
earlier section. Actually the nuclear scattering contribution to

this reflection is even smaller than is shown in Fig. 8 because there is some second-order wavelength contamination from other reflections which accidentally contribute intensity to the position of measurement. In contrast to this distribution of intensity in the (111) reflection, Fig. 9 shows the intensity variation with field in the (220) reflection. For this reflection, the nuclear and magnetic contributions are about equal so that the intensity splitting

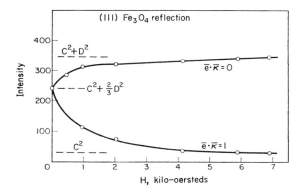

FIG. 8. Magnetic saturation for the (111) Fe_3O_4 reflection for different field directions. The variation with field strength illustrates the realignment of magnetization within the various domains into a direction parallel to the field. The nuclear scattering C^2 is seen to be very much smaller than the magnetic scattering for this reflection.

is not as pronounced as for the (111) reflection. This close equality of nuclear and magnetic scattering in the (220) reflection allows for very interesting neutron polarization phenomena which will be discussed in a following section.

In all of the foregoing considerations, the agreement between experiment and theory was based upon the Schwinger [12] $(H-J)$ formulation of the interaction vector $\mathbf{q}$. The numerical value of this vector squared follows from Eq. (2) as

$$q^2 = \sin^2\alpha, \tag{9}$$

[12] J. S. Schwinger, *Phys. Rev.* **51,** 544 (1937).

where α is the angle between the scattering and magnetization vectors, e and κ. Prior to the Schwinger and $(H-J)$ treatments, Bloch[13] had developed the interaction vector into the form

$$q_0 = e(e \cdot \kappa) \qquad (10)$$

with $q_0^2 = \cos^2\alpha$. The difference between these two expressions for the angular dependence of magnetic scattering is related to

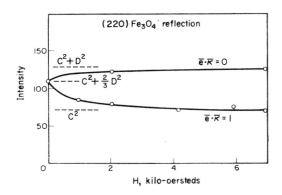

FIG. 9. Magnetic saturation for the (220) Fe_3O_4 reflection for different field directions. For this reflection the nuclear and magnetic scattering components are about equal.

the question as to whether the neutron in its passage through a ferromagnetic medium is affected by the magnetic field strength H or by the magnetic induction B. Ekstein has discussed this relationship recently[14] with particular reference to the interpretation of magnetized mirror experiments. The present experiments on the effect of magnetization direction upon the scattered intensity serves as a direct confirmation of the form of the magnetic interaction. After correcting the (111) Fe_3O_4 reflection for residual nuclear and second-order wavelength scattering, the remaining

[13]F. Bloch, *Phys. Rev.* **50**, 259 (1936); **51**, 994 (1937).
[14]H. Ekstein, *Phys. Rev.* **76**, 1328 (1949).

magnetic scattering cross section has been interpreted in terms of q^2 for four different experimentally-used values of the angle α: $0°$, $23°$, $54°$ (effective value for unmagnetized sample), and $90°$. These values for q^2, normalized to the unmagnetized value of $\frac{2}{3}$, are shown in Fig. 10 along with the two theoretical curves. Within the experimental uncertainty of perhaps 2 percent the agreement with the Schwinger–Halpern–Johnson treatment is very satisfactory. Hughes and Burgy[3] in a recent study of critical

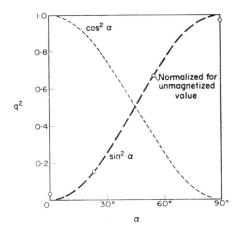

FIG. 10. Variation of q^2 with the angle α between the magnetization and scattering vectors.

reflection from magnetized mirrors have arrived at a similar conclusion although perhaps not within as close limits as in the present work.

Polarization effects in the neutrons scattered by magnetized samples

The commonly used method of producing a polarized neutron beam has been by transmission through highly magnetized samples of a ferromagnetic substance, usually pure iron, which possesses a different transmission cross section for the two characteristic spin states of the neutron. Supplementing this technique, Hughes and Burgy[3] have reported on the polarization effects associated with highly collimated neutron beams reflected from magnetized mirror surfaces. The present experiments are concerned with polarization of the neutrons diffracted by a crystal and are thus complimentary to the usual transmission experiments.

In the case of polarization by transmission, neutrons of one spin state are scattered out of the direct beam preferentially to those of the opposite spin state and a high degree of polarization is reached only asymptotically as larger and larger thicknesses are traversed. When one observes directly the neutrons diffracted by a small crystal it is theoretically possible to obtain complete polarization in a single scattering process. This follows from Eq. (1) in which, for the case of magnetization perpendicular to the scattering vector, $q^2 = 1$ and the differential cross section for unpolarized incident neutrons reduces to $(C \pm D)^2$. This is evident from the fact that an unpolarized beam can be resolved into two equal and independent components of opposite spin state which can be arbitrarily taken parallel and antiparallel to the direction of the applied field. If the magnetic amplitude D and the nuclear amplitude C are equal for some reflection (hkl) of a crystal, the scattering will be zero for one spin state and the diffracted beam will be completely polarized. Lack of equality in the amplitudes will result in only partial polarization. As seen from Fig. 3 the first reflection from cobalt satisfies the condition for nearly complete polarization.

For these cases then where the nuclear and magnetic scattering amplitudes are known, it is possible to predict for a polycrystalline sample the degree of polarization for any given reflection and for any direction of observation relative to the direction of magnetization. The intensity of the diffracted beam from a powdered crystal is very low and hence it was considered of practical interest to see what degree of polarization could be achieved in the much more intense reflections from single crystals. The relation between the diffracted intensity and the atomic scattering amplitudes for a real single crystal depends strongly on the mosaic structure of the crystal and hence it is necessary to actually measure the polarization which can be obtained in a given case.

The polarization in the reflected radiation has been studied in two cases: the (110) reflection for Fe and the (220) reflection in Fe_3O_4. By calculation the maximum polarization to be expected in the former case is about 60 percent whereas it should be 100 percent in the latter case because of the close equivalence of the amplitudes as is evident in Fig. 9. The iron single crystal was prepared by the Bridgeman method from the melt by Mr. Frank Sherrill of this laboratory and contained 5 weight percent of silicon for phase stabilization. A thin slice of metal about $\frac{1}{2}$-in. round and $\frac{1}{8}$-in. thick was cut from the single crystal ingot with the Fe (110) plane in the surface of the slice. A natural crystal of magnetite, Fe_3O_4 of unknown purity, was used in the second examination again with a thin slice about $\frac{1}{16}$-in. thick cut out along the (220) plane. Magnetization of the polarizing crystals was accomplished with an Alnico permanent magnet with H at the crystal surface about 4500 oersteds. The magnetization direction was within the plane of the crystal slices and along the [100] direction in both cases. Neutrons were reflected from internal planes within the crystal slice, the crystal being set in transmission orientation. A schematic diagram of the experimental arrangement is shown in Fig. 11.

The degree of polarization in the Bragg reflected neutron beam was studied by passing it through magnetized analyzing blocks of pure iron. Since the transmission cross section of the magnetized, analyzing block will depend upon the polarization, the latter can be evaluated if the properties of the analyzer are known. The analyzing block thickness was normally 1 in., although measurements were made with other thickness, and it was magnetized with an electromagnet to a field strength of about 8000 oersteds. The experimental procedure consisted of measuring the intensity of the polarized beam after passage through the analyzer when the latter was magnetized parallel to the polarizing fields and when unmagnetized. A magnetic field survey was conducted of the stray field between the polarizing and analyzing magnets and it was established that there was always a residual field of at least one hundred oesteds parallel to these two source fields. This situation sufficed then to insure that the neutron polarization was not altered in passing from polarizing region to analyzing region. Reversing the analyzing field with consequent depolarization effect on the beam was found to seriously effect the transmitted intensity as was to be expected.

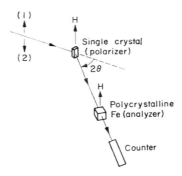

FIG. 11. Diagram of the experimental arrangement used in the study of neutron polarization produced by Bragg scattering from a magnetized crystal.

Halpern and Holstein[1] have given general formulas describing the passage of a neutron beam through magnetized media, including the beam depolarization effects which arise when the medium is not magnetically saturated. Since the analyzing magnet in the present study was not of sufficient strength to saturate the analyzing block, it was necessary to take into consideration these depolarization effects. Their equations show that the depolarization can be allowed for in the polarization evaluation if measurements of the single transmission effect in the analyzing block are obtained with both the polarized beam and with an unpolarized beam. The latter was obtained by substituting a Cu single crystal in (111) reflecting orientation for the polarizing, ferromagnetic crystal.

In this method of polarization analysis, the nuclear and magnetic scattering amplitudes for Fe as determined in the earlier studies were used. Also in the theory application it was

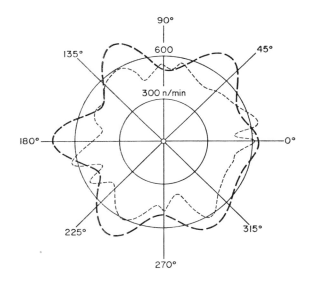

FIG. 12. Intensity variation around the (110) Fe Debye ring for two typical analyzing blocks of thickness about one centimeter. The dashed line is for a sample of cold-rolled steel (grain size about 25μ) and the dotted line is for an Armco block (grain size about 250μ).

assumed that there existed no preferred orientation of crystallites in the polycrystalline analyzer and that the grain size was sufficiently small that the Debye-Scherrer diffraction rings were uniformly populated. This assumption, which has been implicitly accepted in all transmission studies to date, was found to be rather difficult to meet and is worthy of discussion. In studying the polarization produced in the (110) reflection from the magnetized

Fe crystal, discordant results were obtained for various analyzing blocks cut from cold-rolled steel and Armco iron and further study showed that this was due to strong preferred orientation in the former and to large grain size (250μ) in the latter. As typical of these effects, there is shown in Fig. 12 the intensity distribution around the (110) Debye rings produced by two of the analyzing blocks with monochromatic incident neutron radiation. In all of the theoretical formulas developed for the polarization cross section it is assumed that this intensity distribution is uniform so that it is not surprising that the theory application would be poor for such an analyzer. Hughes and co-workers[4] have already pointed out that single transmission measurements are affected by orientation in the block although they do not appear to have taken it into account in their evaluation* of the polarization cross section for a thermal neutron distribution. It was found possible to obtain essentially complete randomness in grain orientation with a suitably small grain size (about 10μ) by solidification from the melt followed by repeated quenching of hot Armco blocks (900°C) in a dry ice-acetone mixture. This material was then cut into suitable analyzing blocks and these were used for studying the polarization in the Bragg beam from a polarizing crystal.

Table III summarizes measurements taken of the single transmission effect in a 1-in. thick analyzing block for the three different Bragg reflected beams. The unpolarized beam from a Cu crystal showed a $+4\cdot6$ percent single transmission effect in the analyzing block and this served to establish the depolarization parameter in the analyzing block arising from incomplete magnetic saturation. Using the observed $-3\cdot0$ percent effect with the polarized Fe beam, the polarization is calculated to be about 41 percent. This is somewhat smaller than the theoretical value for

---

*This may have contributed somewhat to the differences between the Stanford and Argonne Laboratory determinations of the polarization cross section for a thermal neutron beam (see reference 2). A large part of the difference appears to have risen because of differences in the thermal neutron spectrum.

Fe (110) of 63 percent and may be caused by crystal extinction effects, silicon impurity in the crystal or lack of magnetic saturation in the polarizing crystal. All of these effects would tend to reduce the polarization from that predicted by the simple theory. The observed single transmission effect for the Fe_3O_4 (220) polarized beam is very much larger than for the other reflections, amounting to +24 percent and the evaluated polarization turns out to be 104 percent, which of course is impossibly high. It is felt that the measurements are accurate enough to say that the polarization is complete within about 5 percent which means that the relative neutron population in the two spin states is in ratio at least 40 to 1.

TABLE III. COMPARISON BETWEEN EXPERIMENTAL AND THEORETICAL
NEUTRON POLARIZATION IN Fe (110) AND Fe_3O_4 (220) REFLECTIONS

	Single transmission effect in analyzing block (1 in. thick)	Polarization calculated from data	Theoretical polarization for an ideally mosaic crystal
Unpolarized			
Cu (111)	$+0.046 \pm 0.002$	—	—
Polarized			
Fe (110)	-0.030 ± 0.005	41%	63%
Fe_3O_4 (220)	$+0.24 \pm 0.010$	104%	100%

A very interesting further observation is to be noted in the data of Table III and this is the fact that the signs of the single transmission effect in the analyzing block for the Fe (110) and Fe_3O_4 (220) reflections are reversed. This indicates that the directions of polarization in the two polarized beams are opposite and hence that the atomic magnetic moments which are responsible for the polarization are directed oppositely with reference to the applied magnetic field on the two crystals. This is just the conclusion to be drawn from an analysis of the two magnetic structures. In the iron crystal all of the magnetic moments are aligned parallel to the applied field direction. In the magnetite crystal, on the other hand, the octahedral magnetic ions are aligned parallel to the field since they are more abundant than the antiferromagnetically coupled tetrahedral ions so that the latter are aligned antiparallel to the field. According to the crystal structure factors for magnetite given in Table II, only the tetrahedral ions contribute to the (220) reflection and hence the reversed polarization in this reflection is to be expected. The polarization observation for Fe_3O_4 can be considered a very direct proof of the antiferromagnetic nature of the magnetite structure. Interestingly there are other Fe_3O_4 reflections such as the (400) which should produce the conventional polarization sense as characterized by all of the metallic iron reflections.

It is possible to obtain approximately 10^5 monoenergetic neutrons per second in the (220) Fe_3O_4 reflection using neutron radiation from a pile with a central flux of 10^{12} neutrons/cm² sec. This compares very favorably in intensity with other methods of polarized beam production mentioned briefly at the beginning of this section. With regard to the three methods of polarization, viz., (1) transmission through polycrystalline iron, (2) critical reflection from magnetized mirrors, and (3) Bragg scattering from ferromagnetic crystals, there appear to be unique characteristics for each and the selection of the optimum method should be decided by the requirements of the use to which the polarized beam is to be put.

Acknowledgments

We wish to express our indebtedness to Dr. L. C. Biedenharn and Dr. S. Bernstein for helpful discussions and to Dr. M. K. Wilkinson and Mr. L. A. Rayburn for assistance during the polarization measurements.

Index of Subjects
Discussed in the Commentary